Peter Martin worked as a professional civil engineer for over 50 years starting in the days when calculations were carried out with the aid of slide rules and 7-figure log tables. During his career, he designed and supervised construction of many bridges and harbour works in the UK and throughout the Far East, SE Asia, and Africa, living in the East with his family for 12 years. He has three grown-up sons and six grandchildren and now lives in a village near Glasgow where he attempts to keep the garden in some sort of order.

To my sons David, John and Stephen. Also, to the memory of Margo, my wife of 45 years, and my companion and friend who shared so much with me.

Peter Martin

HOME AND AWAY
A CIVIL ENGINEERING ODYSSEY

AUSTIN MACAULEY PUBLISHERS™
LONDON • CAMBRIDGE • NEW YORK • SHARJAH

Copyright © Peter Martin 2023

The right of Peter Martin to be identified as the author of this work has been asserted by the author in accordance with sections 77 and 78 of the Copyright, Designs and Patents Act 1988.

All rights reserved. No part of this publication may be reproduced, stored in a retrieval system, or transmitted in any form or by any means, electronic, mechanical, photocopying, recording, or otherwise, without the prior permission of the publishers.

Any person who commits any unauthorised act in relation to this publication may be liable to criminal prosecution and civil claims for damages.

The story, the experiences, and the words are the author's alone.

A CIP catalogue record for this title is available from the British Library.

ISBN 9781398478107 (Paperback)
ISBN 9781398478114 (Hardback)
ISBN 9781398478121 (ePub e-book)

www.austinmacauley.com

First Published 2023
Austin Macauley Publishers Ltd ®
1 Canada Square
Canary Wharf
London
E14 5AA

First, my thanks to the host of engineers I encountered in my working life: in firms of consultants and contractors and in client organisations in both private and public sectors. Most were a pleasure to work with, whether on "my side of the table" as colleagues, or on the other side in the cut and thrust of construction contracts, and with many of them there was often a pint or two to be enjoyed after work. Without all those engineers and the unnumbered site foremen, plant operators and labourers who make construction works possible there would be no memoir to write.

Thanks are due to Andy Walker of NIRAS Fraenkel who provided photographs from the NIRAS archives and to John Wallace, formerly Chief Executive of Peterhead Port Authority who gave permission for me to use the Peterhead Harbour photographs, initially for an ICE technical paper in 2011 and which I have now included in the memoir, and to Andrew Inkster of Shetland Islands Council for photographs of the Sullom Voe oil terminal.

The memoir is written from personal recollection I am indebted to friends and colleagues who helped clarify some of the details of the projects I have included in it, particularly Robin Theobald, long-term colleague in the Fraenkel organisation on issues with the Tilbury Docks floodgate, Mark Cheung of Mannings in Hong Kong for checking local place names and Cantonese usage generally, and Patrick Augustin of PFAA for similar help with people and places in Malaysia. Any errors in descriptions of projects or translation which remain in the text are entirely of my making.

Finally, I am grateful to the South China Post for permission to quote the amusing poem by "Templar" in the publication "Land of Hope" which aptly describes an aspect of Hong Kong life in the 1980s.

Table of Contents

Abbreviations   11

Glossary   15

Foreword   19

Introduction   21

Part 1   25
   University and First Steps   27
   Getting the Knees Brown   62
   Eastern Suns and the Autumn Moon   109

Part 2   169
   London   171
   The Other Side of the Coin   185

Epilogue   225

Photographs   236

Abbreviations

ADB – Asian Development Bank

AfDB – African Development Bank

AWA – Anglian Water Authority

CBE – Cleveland Bridge & Engineering Company

CEDD – Civil Engineering & Development Department

CPT – Cone Penetration Test

CT – Communist Terrorist (Malaya 1948-1960)

DCM – Deep Cement Mixing for ground improvement

E&M – Electrical & Mechanical

EPD – Environmental Protection Department (Hong Kong)

ESF – English Schools Foundation (Hong Kong)

FE – Finite Element (analysis)

FPSO – Floating Production Storage and Offloading Unit

FFP – Freeman Fox & Partners

GCO – Geotechnical Control Office (Hong Kong)

GF – General Foreman

HA – Normal loading specification for bridges (UK)

HB – Abnormal loading specification for bridges (UK)

HD – Housing Department (Hong Kong)

HKHA – Hong Kong Housing Authority

HyD – Highways Department (Hong Kong)

ICE – Institution of Civil Engineers

IOW – Inspector of Works

IStructE – Institution of Structural Engineers

JKR – *Jabatan Kerja Raya* (Malaysian PWD)

KCC – Kowloon Cricket Club

KCR – Kowloon-Canton Railway

KMB – Kowloon Motor Bus

MTR – Mass Transit Railway Corporation

NGO – Non-Governmental Organisation

NT – New Territories (Hong Kong)

NTDD – New Territories Development Department

OTC – Officers Training Corps

PBA – Peterhead Bay Authority

PD – Principal Datum (for levels in Hong Kong)

PFP – Peter Fraenkel & Partners

PII – Professional Indemnity Insurance

PKK – *Partiah Karkeran Kurdistan* (Kurdish Workers Party)

PLA – People's Liberation Army

PLA – Port of London Authority

POT – Port of Tilbury

PPA – Peterhead Port Authority

RPG – Rocket propelled grenade

SDD – Scottish Development Department

SI – Site investigation

SIC – Shetland Islands Council

SVT – Sullom Voe Terminal

UMNO – United Malays National Organisation

USRC – United Services Recreation Club

Glossary

Ah – *(Cantonese)* prefix to surname of favoured employee

Batu – *(Malay)* stone, rock

Boko Haram – Jihadi terrorist organisation in north-east Nigeria

Booms – Beams or struts connecting pontoons to shore

Buaya – *(Malay)* crocodile

Bukit – *(Malay)* hill

Bumiputra – *(Malay)* Indigenous people, lit. son of the soil

Catty – Unit of weight in Hong Kong and Malaysia, abt. 1.3 lbs

Chau – *(Cantonese)* island

Chu – *(Dzongkha)* river

Chunam – *(Tamil)* Impermeable plaster on surface of slopes

Dai pai dong – *(Cantonese)* food stall (Hong Kong)

Dzong – *(Dzongkha)* fortress, administrative centre of region

Fung shui – Chinese geomancy, lit. wind-water in English

Gabion – Rock filled cage often used in civil engineering

Gazette – Official publication of the Government of Hong Kong

Godown – Hong Kong, Singapore: warehouse

Goh – *(Cantonese)* classifier usually with numbers of nouns

Gweilo – *(Cantonese)* pale ghost or foreign devil, derogatory

Istana – *(Malay)* palace, residence of Sultan or King

Kedai – *(Malay)* small restaurant, lit. shop

Klong – *(Thai)* canal or waterway, particularly in Bangkok

Kuomintang – Chinese Nationalist Party

Hantu – *(Malay)* ghost

M'goi – *(Cantonese)* thank you, please, excuse-me

Mama san – Woman in position of authority (Japan and Far East)

Mentri Besar – *(Malay)* Chief Minister

Naira – Currency of Nigeria

Nullah – Open concrete lined drainage channel (Hong Kong)

Orang – *(Malay)* man (e.g. *orang Japan*, Japanese man)

Padang – *(Malay)* open recreational space in town, lit. field

Parang – *(Malay)* large, curved bush knife

Penanggalan – *(Malay)* a particularly malevolent ghost

Perunding – *(Malay)* consultant

Peshmerga – *(Kurdish)* freedom fighters, lit. those who face death

Pintle – Pin or boss on which something turns

Shakti – *(Dzongkha)* goddess in form of Buddhism in Bhutan

Sungai – *(Malay)* river

Tanjung – *(Malay)* cape, point

Shan – *(Mandarin and Cantonese)* mountain

Wan – *(Cantonese)* bay

Yum cha – *(Cantonese)* brunch with tea and dim sum, lit. drink tea

Foreword

The memoir covers my working life as a civil engineer from university in the 1960s to retirement over 50 years later; a career spent mainly with one firm of consulting engineers. It draws on personal reminiscences from early days on projects in Scotland, through work overseas, particularly in Hong Kong and SE Asia, to the latter part of my career as a partner and managing director of the firm in London and Surrey. As well as focusing on some of the larger projects which I became involved in it describes political developments and background events which sometimes drove the need for them. Although of necessity some technical descriptions are included these have been kept to the minimum and the use of specialist engineering terms has been avoided as far as possible.

In the first part of the memoir the projects described are largely in chronological order. They chart the rapid rise of the firm I joined from its beginnings with a small number of staff to a well-regarded medium-sized firm of consulting engineers responsible for prestigious projects across the world. In parallel with the expansion of the firm, I took on increasing individual responsibility for projects the firm had been awarded, sometimes close to the limits of my experience. For much of that time there was an exhilaration in this parallel growth of the firm and my own abilities as a civil engineer. This was enhanced for me by the undoubted pleasures of expatriate life in the Far East where I was based for twelve years and some of these are touched on in the memoir.

The second part describes a less happy situation, initially at least. I had returned to the UK to take up a role as director of the company when I found it to be in a parlous financial position close to liquidation as a result of serious mismanagement by other recently appointed directors. This was narrowly avoided by a partial merger with a newly privatised organisation and with a struggle the core business which I remained in was built up again on a more rational basis. Old clients that had been lost were regained and interesting assignments were obtained. Some of the latter projects for those clients along with others of interest are described, including the design of a large project back in Hong Kong at the very end of my career.

The memoir concludes with personal views on the troubles the firm suffered, and how the civil engineering profession organises itself. The future of civil engineering in a rapidly changing world is also considered, and finally, I have attempted to put into perspective my career as a civil engineer.

Peter Martin
February 2023

Introduction

An engineer is a man who can do for five bob what any damned fool can do for a quid.   *– Unattributed. Quoted by Neville Shute in Trustee from the Toolroom*

On a bright and breezy morning in late August 1968, I parked my car at Carradale Harbour on the west coast of Scotland and walked across the road to some huts on a grassy strip at the shore. Beside the huts, there were stacks of black painted steel frames and sheet piles and next to them a tracked crane which I later learned was an RB 22, the RB standing for Ruston Bucyrus and the workhorse of small-scale civil engineering construction in the 1960s. It was eight o'clock and I was about to start on site as Resident Engineer on an extension to the breakwater and boat shelter in Carradale Harbour.

The door of the end hut was open revealing hand tools and other equipment; it was clearly a store. There were voices coming from the next hut in line and I knocked on the door and went in. Five men were sitting at a table having a breakfast of bacon sandwiches and drinking tea. One stood up and held out his hand which I shook. I was expected. He was the contractor's general foreman or GF—the man who would run this small job on site for the contractor. He had bright blue eyes, greying hair and a pleasant Scottish accent which I could not immediately identify, and he introduced me to the others, a crane driver and three labourers.

I was invited to sit down and join them, which I did, declining the bacon as I had already eaten, but taking a mug of tea. After I had drunk the tea, the foreman took me to the third hut which was to be mine and handed me the key. There was a desk and a drawing table, some chairs, a cupboard with a waterproof jacket in it along with wellingtons of different sizes and a telephone which the GF said was not yet connected. He left me to survey my small domain. I had spent four years at university studying civil engineering and two years working with a structural firm, but I felt this was the real start of my career as a civil engineer, and I realised how little I knew.

I cannot say I had a burning conviction when I was young that I was a born engineer, and that no other vocation would do for me. Up until around 1960,

National Service (conscription) was still in place in Britain as it had been since the Second World War. As my schooldays were coming to an end and as I thought I would spend two years in one of the armed services before going on to some sort of further education, there seemed to be no urgency in deciding on a career. In common with others in Glasgow High School, I had joined the Officers Training Corps in my second year as there was a belief amongst the pupils in the school that those who had been in the OTC were more likely to be selected for officer training when called up.

As it turned out, once in the Corps as we called it, I found I had some aptitude for weapon training and the other military disciplines we learned and I thought an army career, or at least a short service commission, were possibilities for me. But when I was entering my fifth year at school in 1961, National Service was being phased out. There was no longer any requirement for military service and the idea of an army life became less appealing.

This change of heart did not immediately lead to ideas of other options for a career. Although time was moving on, I gave little thought to what I might do when I left school, other than probably go on to university in some suitable course. I was quite enjoying school life with sport and the Corps and getting out and about with friends to pubs and parties in the Glasgow suburbs. I was becoming reasonably competent at maths and physics, I enjoyed English and surprisingly, even French literature which I was introduced to by one of our French teachers. In truth, I was probably immature and not in any hurry to leave the comfortable and sheltered life of school and home now that I would not be compelled to by conscription.

But my father had always said I should become a civil engineer. Before and during the Second World War, he had been a deck officer in the merchant navy, sailing around the world, much of which at that time was in the old British Empire. In the ports he went to he saw at first hand major civil engineering works in the form of harbours and bridges designed and built by British engineers, especially in India and the Far East, and it seemed to him that the life of those engineers was interesting and worthwhile. It probably was at a time when there was little competition from foreign engineers; only British organisations could get a look in. Civil engineering was not an office-bound job, he said, and civil engineers could expect to be out in the open air a lot in connection with their works. This appealed to me as I always wanted to be out and about myself and gradually, at the back of my mind, I suspected that is what I would become, or at least study at university. It seemed as good a career as any.

The recommendation by my father was later reinforced by Major Alistair Duncan Miller M.C. of Remony estate in Perthshire, "the Laird", for whom my uncle Johnny worked as the head gamekeeper on the estate. The Laird had been in the Royal Engineers during the second world war and had won his M.C. in the NW Europe campaign in February 1945 in a bridging operation under fire, as the British army advanced through the Low Countries towards the borders of Germany. After the war he was Resident Engineer for Gibb on the Ben Lawers hydroelectric scheme before he inherited the estate on the death of the old Laird in 1949.

Every summer, as a boy, I spent several weeks staying with my uncle in the keeper's house at Remony where I learned to fish and shoot (once successfully stalking a stag and being "blooded" by my uncle) and had some elementary riding lessons from the Laird's wife, Mrs. Miller. When my father came up from Glasgow, he would sometimes fish on Loch Tay with the Laird. At some point he must have mentioned his ideas for my career and the Laird took this up with me on several occasions when I was helping with the horses at shooting parties on the grouse moor (at one pound a day), pointing out the advantages of becoming a civil engineer. With such advice I could scarcely consider any other career: civil engineering it would be.

There was nothing remarkable in what I did in my work. While I was personally responsible for the design of interesting (to me at any rate) and enduring bridges and harbour works in various parts of the world and later directed staff under me on some significant projects, many other engineers of my generation could claim the same thing, and often for works considerably more prestigious than mine. Nevertheless, I can take some satisfaction in *having directed the forces of power in nature for the use and convenience of man* as the 1828 definition of civil as distinct from military engineering put it. I enjoyed what I did, in the technical sense at least, although not the financial worries in making ends meet as a partner in a small independent engineering consultancy, sometimes possibly punching above its weight in an unforgiving market. If I did achieve anything worthwhile, of course, it would not have been possible without my wife beside me supporting and advising me and taking care of our family issues over the years.

In describing the civil engineering works I was involved in, it has sometimes been necessary for better understanding of why the projects were constructed, to describe background events in the wider world which may have prompted them. My description of these events may be coloured by my own views and prejudices, although I have tried to be as objective as possible.

My early years at university and in various firms of consulting engineers, working in design offices and on sites, seem so remote from my life now that they might have happened to someone else. Did I descend 15 metres into unlined pile holes only one metre in diameter through boulder clay in Glasgow with my feet in a bucket lowered from a crane, and again do the same thing in Hong Kong in hand-dug caissons of the same small diameter? Did I walk without any harness or other safety measures on the bare, partially erected steelwork of Ballachulish Bridge in Scotland, 25 metres up with the waters of the Narrows sweeping underneath me on the 7knot ebb tide and buffeting wind gusts threatening to blow me off the girders?

These things are inconceivable today in our safety-conscious world. Or did I go down 20 metres in a diving bell in the muddy waters of the Thames in the Western Entrance Lock at Tilbury Docks to probe the extent of a scour hole (or possibly a WW2 bomb crater) close to the proposed foundations of the floodgate we were to build there? And later, was I really shot at in an ambush in Kurdistan in Northern Iraq by Saddam Hussein's people, fortunately well protected by Kurdish *peshmerga* in our vehicles, who killed or disarmed the ambushers so that when we travelled on to the bridge I was to inspect, I was holding a loaded Kalashnikov AK-47 assault rifle taken from one of them? It seems I did all those things as I have a few old drawings and correspondence files to prove it.

I kept no diaries and must rely mostly on imperfect memory for the thoughts I had and the names and actions of colleagues as projects and situations unfolded. Inevitably there will be inaccuracies and my research into outside sources may sometimes be insufficient, but I hope to present a picture of some of the highlights of my career and bring out such satisfaction as it brought me, as well as its downsides.

Part 1

University and First Steps

A little learning is a dangerous thing,
Drink deep, or taste not the Pierian spring,
There shallow draughts intoxicate the brain,
And drinking largely sobers us again.
From Essay on Criticism, Alexander Pope

I left school at the end of June 1962, and went to my last Corps Camp, at Cultybraggan Army Training Camp near Comrie in Perthshire. As we were told, Cultybraggan had been built in 1941 as POW Camp No. 21 to house the most committed and fanatical Nazi POWs. After the 10-day camp, during which I had finally obtained the coveted Marksman Badge on the firing range, too late unfortunately to flaunt it at school, I spent the rest of the summer driving a van delivering machine tools around Glasgow for my father's firm. The term at Glasgow University started in early October and I went through the time-honoured process of matriculation, freshers' nights and acquainting myself with university life and the pleasures of the Beer Bar. I lived at home as did most of my classmates who were predominately from the Glasgow area with only a scattering of students from England and abroad who stayed in Halls of Residence in their first year.

Apart from laboratory work on some afternoons, we sat at desks much as we had done in school, but the teaching was a great deal more impersonal than I was used to. You were on your own and whether you sank or swam was up to you and of no great interest to the lecturers whose job, for first year students at least, was to deliver the lectures and course work and nothing more. In Glasgow University the course work in Years 1 and 2 was common to all engineering students whether they would go on to civil, electrical, mechanical or aeronautical engineering. It therefore covered a lot of ground. Everything was much more difficult than I had imagined, and it became clear the Scottish students were behind our English counterparts in mathematics, physics and chemistry. The more specialised English A-Levels in these subjects were far more relevant to the technical teaching we were getting than our

Highers which had been obtained in the more broadly based Scottish education system. It was almost as if we were a year behind in the sciences. At the time English university engineering courses were only three years long compared with our four at Glasgow and the technical subjects in the A-Levels were geared to this. So far ahead were they that the English students could probably have gone straight into our second year, while we certainly needed that first year just to catch up.

Having said this, however, much later as an employer, I came to realise that the great strength of the Scottish system in my time at least was its broad base. We had continued with English right up to the fifth year in school, covering both grammar and literature which are so important in communication of ideas, whereas most of the English students had dropped English after sitting their O levels. I found communication skills lacking in many English graduates and other things being equal, I tended to recruit new graduates from Scottish universities more on the strength of their facility in English than anything else. This owed nothing to nationalistic favouritism and was rooted in commercial reality. Sadly, successive Labour and SNP governments in Scotland have progressively dumbed down the education system in their time in power and I doubt if my observations hold good today.

I struggled with the course in the first year and only just scraped through the end-of-term degree exams. The second year was no better. I had great difficulty with maths, grappling with the complexities of Fourier series, third order differential equations and imaginary numbers with their negative squares which I had previously understood not to be possible. It appeared I had reached the limit of my ability as far as maths was concerned. Thermodynamics was another problem subject for me along with electricity which, of course, made use of a lot of the maths. There were six separate degree exams at the end of the second year and I failed three of them to be faced with resits which I would have to pass to stay in the course. I did not take a summer job that year and buckled down to study each of the failed subjects, studying harder than I had ever done before and since. Somehow, I passed all the resits in September and thus advanced to the third year, but I dropped into the Ordinary Class in the process. This was not a big issue in those days as most engineering students at Glasgow took an Ordinary Degree with only a small group going for Honours, unlike today when Firsts are handed out with abandon.

Suddenly, life became easier. I had broken through from the theoretical engineering course into the study of civil engineering pure and simple and no longer had to worry about high level maths and complexities of electrical theories. It was not that the civil engineering subjects were easy, but they were of more interest to me than imaginary numbers and so on and seemed more practical so that one could see the point of them in the real world. I discovered I liked structural analysis, and this influenced me through my subsequent working life and ultimately directed my career towards bridges and harbour infrastructure. Suddenly, the lectures and course work became enjoyable. I respected both civil engineering professors, the Regius Professor William Marshall, a patrician figure who was famous for structural developments in plastic design of steelwork during the war while in Imperial College and Professor Hugh Sutherland, more down to earth, perhaps literally, as he specialised in soil mechanics.

The lecturers under them were excellent as well, particularly James Lindsay on design of concrete structures and Hugh Nelson on steelwork. They took time to explain and guide us through the various design methods in these materials, unlike the remote and somewhat unhelpful university teachers of the first year. In this environment my third and fourth years passed quickly and I had little difficulty with the final examinations at the end of the fourth year. Now I had to move on into engineering employment in the real world.

In 1966, when I graduated jobs were abundant and representatives of consulting engineer and contractor organisations came to the university giving presentations to final year students on the merits of their firms in attempts to woo us into joining them. The aim of most of the final year students was to get into a firm with a recognised training scheme leading to membership of the ICE. For this it did not really matter too much whether one joined a contractor or a consultant as a requirement of the ICE was to have had both site and design office experience before applying for the membership interview, usually an absolute minimum of three years of experience between the two. However, it was probably easier to get the experience with consultants as they could usually provide it both in their design offices and on sites in supervision of works, while contractors might not be able to provide design office experience so easily without farming out the trainee to a consultant, which they might be reluctant to do if they had a lot of work on their books. It was also true that, while contractors paid more to the trainees than did consultants, they expected them to work much longer hours on site in return.

I decided to go down the consultant route and narrowed down the choice of employer to just two of the organisations who had made me offers. The first was from the well-known civil engineering consultant Scott Wilson Kirkpatrick to work in their London office and the second was from a local structural firm, T. Harley Haddow & Partners for their Glasgow office. I chose Harley Haddow on the grounds they were paying slightly more than Scott Wilson, and I was not sure I could survive in London on the salary Scott Wilson had offered. In retrospect this was probably a mistake in career terms and Professor Sutherland obviously thought that way while not saying so in so many words, as he expressed some disappointment when I told him. He said that I would undoubtedly gain a lot of experience in reinforced concrete with the structural firm, but I would be out of the mainstream of civil engineering.

He was right and I just marked time in the two years I was with Harley Haddow. Their Glasgow office was a pleasant working environment with two decent partners, and I was recruited to play football for the head office in Edinburgh. I made plenty friends in the staff, and I did learn to design reinforced concrete structures, but not much beyond that. I found it frustrating working on projects in which architects took the lead role and made sure I rarely worked with them again for the rest of my career, other than to employ them in a sub-consultant capacity. After two years I decided I had better get into real civil engineering. I thought I had probably gained enough design office experience for the ICE membership interview and looked to move to a post which offered time on site. For this, I took a job as Resident Engineer for the construction of a breakwater at Carradale in Argyll for A. M. Robertson, the Engineer on the contract, who had designed the breakwater. He was a consulting engineer based in Helensburgh having been a civil engineer on the staff of Argyll County Council in Lochgilphead before striking out on his own as a consultant. This had been a good move for him as he now took on all the marine civil projects for the council under the generous standard fee conditions which applied to consultants at that time.

Just before leaving Harley Haddow, I took holidays which were due to me and went to France with two friends. We travelled to London and on to Dover to catch a ferry to Calais from where we took the train to Paris, arriving at the Gard du Nord early in the morning. Paris was not looking its best that grey July day. It was only a few weeks after the anti-capitalist student riots in May of that year which had started in the Sorbonne and quickly spread with workers in many industries joining in until the French economy was on the verge of collapse. Leftist graffiti was evident on many buildings and some of the broad boulevards were a mess where cobbles and other paving had been dug up by the students to throw at the police as they baton charged the rioters. The riots had fizzled out in June, but there was still tension in the air in the bistros and estaminets we ate in during our stay. Despite this, I liked Paris and would have stayed longer to spend more to time in the Louvre and other attractions, but as we had only two weeks, we hired a car and moved on along the north of the country to Normandy and Brittany, taking in the magnificent Cathedral at Chartres on our way. We stayed in Brittany with its pleasant beaches for a week, before driving back to Paris and catching the train to Calais followed by the ferry over the Channel again to England. Then it was time to start with Robertson, first for a week in his office in Helensburgh where he briefed me on his design, and then down the long A83 to Carradale.

Carradale Breakwater

Construction of a breakwater is about as far removed from building structures as could be—almost the other end of the civil engineering spectrum in fact. Breakwaters are mainly of two types: rock mounds placed directly on top of the seabed with their sides sloping down through the water and solid vertical sided structures embedded in the seabed. In my subsequent career I was involved in many breakwaters of both types. The function of both of course is to calm the water on the inside of the breakwater and sometimes, with the second type, to provide berths for vessels in the harbour. Carradale Breakwater was of the second type, consisting of parallel lines of steel sheet piles driven into the seabed braced apart by steel frames and steel H-piles with the space between the lines of sheet piles filled with rock for stability. There was an existing short length of breakwater in Carradale harbour, but the extension we built increased shelter from waves considerably and enabled more fishing vessels to be moored on the inside of the breakwater. These vessels were mainly small herring drifters for the fishing grounds in the Kilbrannon Sound between Arran and Kintyre in the Clyde Estuary and the deck of the breakwater was provided with bollards for mooring the vessels and access ladders for the crews. When I started, I was incredibly green and naively thought all I had to do was to ensure the works were constructed according to the drawings. I knew nothing of agreeing monthly measurement of work done with the contractor or preparation of interim payment certificates and did not even understand many of the requirements of the specification for the works.

Fortunately, Robertson was an engineer with wide experience of marine works and his design, although basic, was eminently buildable by an experienced contractor. The contractor for the works was H. M. Murray, a small local contractor ideally suited to such jobs with a good track record of successful marine civil projects throughout Scotland. The pleasant Scottish accent I heard when I met the GF on my first morning was Orcadian; he came from South Ronaldsay, which is the southernmost of the Orkney islands. He was extremely competent and knowledgeable in all marine civil engineering works, and with experienced labourers under him, construction of the breakwater proceeded smoothly without problems of unforeseen conditions of one sort or another which are common on marine projects, the most difficult of all civil engineering works. That 80 metres length of sheet piled breakwater could be built by five men with only a crawler crane and a piling hammer to help them is a measure of their skill. Thinking back on it brings home to me how many unheralded men such as these on sites like Carradale Harbour have constructed facilities people take for granted. Murrays could probably have exploited my ignorance to financial advantage, but never sought to do so and my time on site provided a good introduction to supervision of heavy civil engineering construction, albeit on a small scale. I became friendly with the workers particularly the GF who introduced me to whisky. On a wild, wet day we were all on site till well after dark to finish some part of the works and we ended up at the bar of the Carradale Hotel where the GF insisted on buying me glasses of whisky to go with the pints I was drinking, although I protested that I only drank beer and did not care for whisky. However, it went down well that wet night and I have to say I have drunk the stuff ever since.

When the breakwater contract was completed, Robertson did not have any other site supervision requirements, but he offered me work in his small design office in Helensburgh to which I travelled every day from Glasgow. He was a good engineer, and I learned a lot from him on the way to go about designing marine structures—first and foremost to ensure they were both as robust and buildable as possible. A basic mantra I learned from him and have worked to throughout my career was to so organise the design that most construction operations could be carried out in the dry above water level. This is not always possible, but it certainly pays to minimise underwater operations.

Unfortunately, the A. M. Robertson organisation workload was dropping off through the spring of 1969. I could see I was becoming surplus to requirements and reluctantly decided to move on before I was made redundant. A short-term design vacancy arose with Oscar Faber & Partners in their Glasgow office, and I took that while I looked around for something more interesting. As a firm Oscar Faber & Partners were well known for heavy industrial structures and Faber himself was the author of several good books on this type of civil engineering, one of which I still find very useful. Most of my involvement was in designing structures for a cement works in Dunbar and the large ICI petrochemical plant in Grangemouth under Bill Cantlay who was Oscar Faber's Glasgow office partner. Cantlay was a council member in the Institution of Structural Engineers and suggested I had a shot at their membership examination. This was purely an examination as, unlike the civils, there was no structured training scheme in IStructE, but it was considered extremely difficult with a pass rate of only around 25 percent most years. The exam lasted 7 hours during which time candidates had to complete a preliminary design and specification for a project selected by the candidate from five options in the exam paper. A lot depended on the questions in the paper and whether one had sufficient experience of any of the types of projects in the questions to be able to produce a realistic design in the time available. I applied to sit it with no real expectation of success, but amazingly, I passed. How could this be with my limited experience? It happened that Bill Cantlay was an examiner for the Institution and while he would not have marked my paper, what he had done was to provide one of the projects to be selected by the candidates. This turned out to be a project I had been working on. Everything about it was the same, even down to the ground conditions on the site. I was therefore able to churn out a credible design solution in the allocated time along with a comprehensive specification just as the company had done for the real thing. With this stroke of luck, there I was, ready to become a member of the IStructE, although at 25 years of age I had to wait six months till I was 26 which was the minimum age for membership.

Jedfoot Bridge

Getting into the IStructE did not do anything to help me with membership of the ICE for which I had to submit a detailed description of my training and then, if this was deemed suitable, pass an interview with two examiners grilling me on what I had learned. I knew I was still a short of site experience and moved on again this time as Resident Engineer on Jedfoot Bridge in the Borders for Blyth & Blyth, an Edinburgh firm at that time well known for bridges. This was my first experience of bridge structures which along with marine works were to become the focus of my subsequent working life. Jedfoot Bridge was designed as a single span structure over the river Jed, located about 2km downstream of Jedburgh. Once constructed, along with around 500m of roadworks on either side of the bridge, it would provide a direct route from Jedburgh to Kelso which lay about 8 miles to the east on the A698 from Hawick. The bridge was to be a steel and concrete structure with four steel plate girders acting compositely with a reinforced concrete deck. At the time, this form of construction was gaining in popularity throughout the UK and after going out of fashion for a while has become very popular once more. It has the advantage that the steel girders can be erected quickly over rivers and busy highways with a minimum of temporary supports and once erected, form a ready working platform for constructing a concrete deck.

When I was appointed as Resident Engineer the bridge contract had just been awarded to the civil engineering contractor Tarmac with a start date in the middle of August. Prior to the start I spent a week in the office in Edinburgh getting to know the Blyth & Blyth people in the Bridge Section, particularly Colin Hood who had designed the bridge. I arrived on the site with fields of barley ripening all around at the same time as Tarmac's agent Graham Prince with whom I became friendly. Throughout the autumn of 1968 after clearance of the site, work progressed on construction of approach embankments on either side of the river crossing and installation of the cast-in place concrete piles for the bridge abutments. A cold winter followed with some problems in casting the lower parts of the concrete abutments in sub-zero temperatures, but eventually the abutments were constructed with their bearing shelves for seating the bridge girders. These arrived, each one in three segments as they were too long for road transport and the segments were then welded on site to form full-length girders. They were fabricated by Dorman Long (of Sydney Harbour bridge fame) which by that time had been nationalised and subsumed into the government-owned British Steel Corporation. Once the four girders had been lifted into position by a crane operating on the east bank of the river, steel cross-bracing was bolted between them, formwork was placed between their top flanges and the concrete deck was cast. It only remained to lay hot-rolled asphalt surfacing and fix the crash barriers on each side of the bridge deck. Almost exactly one year after the start the contract was completed.

Working on the Jedfoot Bridge contract made me more comfortable in the role of RE. During the year I spent on site I was able to put together my application and technical submissions for membership of the ICE and attended the interview in the May of 1970. The pass rate for the Civils was around 50 percent at that time and given that I had completed the necessary training and gained what I considered to be good experience I did not expect to have any difficulty. I duly passed, although I was closely questioned by one of the examiners on my knowledge of site welding when I airily mentioned the site butt welds on the main girders of the bridge. This is always a more difficult operation than welding in a closed fabrication shop under controlled conditions and I realised the examiner had picked up on my lack of any real knowledge of the issues involved. Fortunately, I was able to expound more knowledgeably on the problems of winter concreting and this must have saved the day. Once the formalities of the membership process had been completed, I could put the letters MICE after my name along with MIStructE (at the time I obtained my membership, the entry level was as Associate of the Institution rather than Member and strictly speaking it was AMICE, but the designation was changed a year later).

I enjoyed my time in the Borders living in a rented cottage on Newmills Farm close to Jedforest Riverside Park rugby ground. Rugby was and still is a big thing in the Borders, with the annual Melrose Sevens, District games between South of Scotland and other districts and a visit of the Springboks, this held with a large police presence as hundreds of "stop the tour" anti-apartheid protesters were ranged outside the ground. I was even pressganged into turning out at centre for a scratch Jedforest team in borrowed kit against a club team who were touring Scotland and suddenly turned up looking for a game. Fortunately, I had my own football boots as I was playing regularly in office football games in Edinburgh for Blyth & Blyth (and occasionally as a ringer for my old friends in Harley Haddow). I viewed this impromptu rugby encounter with some trepidation, expecting our opponents to have an uncompromising approach, but fortunately they turned out to be even worse than our scratch fifteen and we won by about 20 points. I knew the scrum half as he worked on the site. He had too been coerced into playing he said, having just gone into the club for a drink. He must have been about twenty years older than me, but he ran the show and scored at least two tries himself sniping around the base of the scrum.

Fishing was another interest. Where our road joined the main A698 leading towards Kelso there was a small estate called Mounthooly which had private fishing along a mile of the right bank of the river Teviot. We had done some accommodation works for the owners, who were an elderly couple, improving the entrance to their driveway and installing proper fencing along the road. When this was completed, I went up to the house to ask if I might fish on their stretch of the river. Permission was granted and as I chatted to the old couple, I asked how long they had lived there, to be told the Major had bought Mounthooly just after the war. Looking at them, I guessed this to be the First World War, not the Second and said so. "Oh no. The Boer war" said the lady, putting them well into their eighties. The private water provided excellent fishing and I could generally rely on catching trout in the summer evenings after work. These would be gutted and rolled in oatmeal before being fried up for a late supper, the fish so fresh that the nerves would still be functioning, and the gutted fish would curl up in reaction to the heat of the pan.

In the depth of winter, I saw some fishing of a different sort. The steel fixer on the site told me he had seen some nice salmon in a pool in the Jed close to our site and would bag them that night. Once it was dark, he shone a torch into the pool and a fish swam up to investigate, which he neatly speared with a sharpened reinforcing rod he had prepared. When he got the fish out on the bank, he said it was a hen fish and the cock fish would follow. Sure enough, a few minutes later another fish appeared which he also skewered. We were all eating fresh salmon steaks for the next few nights.

Trondra/Burra Bridges

When the Jedfoot Bridge contract finished Blyth & Blyth did not have any other site work coming up and offered me a job in their design office, but I had already seen a bridge engineer vacancy with W. A. Fairhurst & Partners in their Glasgow office. They had recently been responsible for a new bridge over the Tay at Dundee and were currently designing the Kingston bridge In Glasgow along with all the structures on the western flank of the M8 in the city centre. I applied and got the job which would initially involve working in a team designing a bridge at Ballachulish to replace chain-operated ferries. However, just before I left Blyth & Blyth, at a party in the Station Hotel to mark the completion of the Jedfoot Bridge contract, I was asked by the senior partner in Blyth & Blyth if I could help by undertaking a temporary RE role on the Trondra/Burra bridges contract in Shetland. He said this comprised construction of two identical bridges connecting the islands of Trondra and Burra to the mainland of Shetland. It seemed the regular RE had suffered a serious hernia and would be out of action for a month to six weeks. The name of the contract meant nothing to me as I had never been in Shetland and only knew it was as far north in Scotland as one could go. The bridge partner said he would ask Fairhurst if I could defer starting with them for six weeks. I was a bit miffed at this, but I decided to go along with it and Fairhurst agreed. In the end I was in Shetland for over two and a half months before going to Fairhurst.

In those days Shetland was a much poorer place than it is today. its population had fallen below 20,000 and young people were leaving for the south in increasing numbers. The main occupations were fishing and sheep farming, there being virtually no arable land suitable for growing crops, and at that time neither of these activities made much money. Shetland is now much more affluent, for which the Shetlanders of the present time have their grandfathers to thank, although with a continued involvement in Shetland until quite recently I know that the inhabitants now regard their affluence as a right and give little thought to how it came about.

With a maritime tradition in the islands many of the old Shetlanders had been to sea and had gained a knowledge of the ways of the world on their travels. Large oilfields had just been discovered under the North Sea and Shetland had an ideal deep-water location for an oil terminal with export jetties to connect with the undersea pipelines which would run from the fields. This was in a long sea inlet called Sullom Voe about twenty miles north of Lerwick. In many places the big international oil companies who needed the terminal would have simply come in and bought up the necessary land cheaply from local farmers, but not so in Shetland. It was made clear to the companies by the local council, whose members included many of the astute old seamen, that planning approval would be conditional on a sizeable proportion of the revenue for handling the oil being paid into the community. UK government support for this demand was forthcoming, although the government were always at pains to ensure the terms were not made too punitive in case the companies looked east to Norway for the terminal. The result of the negotiations was the passing of an act of Parliament, The Zetland Act, which set out the broad terms of an agreement for the operation and management of the terminal and the proportion of oil revenues to be distributed to the community. Incredibly, the oil companies, who were to be responsible for the cost of building the terminal, had to hand over ownership of it to the council on completion which is where it remains today with the companies still responsible for annual maintenance. In other words, they were getting a lease of the terminal which they would pay to build and then operate with full maintaining and repairing obligations. Much of the credit for this must go to old council members and their chief executive at the time, a young Englishman called Clark, who led the negotiations. As the royalties eventually flowed into Shetland in large measure, the result was a transformation of the local economy from its poor state in 1970 to one of the richest places in Scotland today. When I went to Shetland in the August of that year, these benefits were still in the future, although the news was full of oilfields being discovered. There was an expectation that the UK could perhaps become less dependent on oil from the Middle East, but the part to be played by Shetland had yet to unfold.

On arrival I visited the permanent RE, Martin Broadgate, in the Gilbert Bain Hospital in Lerwick where he was recovering from his hernia operation. Martin and I were to become very good friends and we worked together on other major projects over the years, in Scotland and Hong Kong. He briefed me on the state of play on the Trondra/Burra bridges contract which he said was going as well as could be expected given the Shetland weather with frequent rain and gales. The contract had been running for a little over a year and site operations had virtually shut down over the previous winter which was sensible given the weather and the short hours of daylight. He expected the same in the coming winter, but he hoped as much progress as possible would be made while I was there before the shutdown. The main contractor for the bridges was Lilley Construction, an old Scottish firm sadly now gone, with Cleveland Bridge & Engineering of Darlington as subcontractor for the bridge steelwork. Shetland has no rivers of any size, just numerous small burns running down the hillsides to the sea. Thus, it needs no large bridges, although there are many small bridges and culverts carrying roads over the burns. However, the bridges I was to supervise were quite large structures about 2 miles apart connecting the two offshore islands of Trondra and Burra to the Shetland mainland. They were both 3 span structures, continuous over piers in the water with central spans of 75m and side spans of 40m. The crossings were from the mainland to the island of Trondra, lying just off Scalloway on the west coast of the Shetland mainland and from Trondra over a narrow, but deep, sea channel to the island of Burra further to the west.

While I was supervising the works in Martin's absence, erection of the Trondra bridge steelwork was carried out. At the Burra crossing the seabed for the deep-water foundations for the bridge piers was excavated and the lower parts of the concrete piers were cast.

The superstructures of the bridges comprised two plate girders braced together which were continuous over the intermediate piers and there was a concrete deck acting compositely with the steel girders like Jedfoot Bridge. The steelwork was erected by launching across the channel with banks of temporary rollers on the abutments and the pier tops. The girders for the Trondra bridge were first assembled on the mainland side from a series of 20m segments joined by high strength friction grip bolted splices using cover plates on the webs and flanges of the girders. They were then winched over the piers by a suitably geared hand winch with lateral guide rollers to keep them in line. The process was very simple and in the main it was accomplished easily. The only real problem was the cover plates on the bottom flanges where the protruding bolts could foul the rollers and as each splice reached the rollers the girders had to be jacked up on to wooden skids before proceeding.

The foundations for the main piers at Burra were cast within permanent pre-assembled steel cylindrical cofferdams which were set in position on the seabed with the aid of divers and cleaned out by grab to remove loose bed material. They were then cast underwater by tremie, a flexible pipe fed by a hopper full of wet concrete with the bottom of the pipe always below the concrete surface in the foundation to ensure no air was trapped within the mass of concrete being built up. All concrete had to be mixed on site as there was no access for ready-mix trucks and the logistics of the operation were critical; mixing large volumes of concrete with a crane of sufficient reach to get the tremie pipe into the cofferdam and hold it there while concrete was fed into the pipe from the hopper which had to be kept charged with concrete at all the time. Mass concrete foundations like these should be cast in one operation where possible to avoid horizontal planes of weakness, but this proved impossible in the circumstances as there was a practical limit to the production rate of site mixed concrete and the hours of daylight were already becoming short in the northern latitudes of Shetland. It seemed we would have to settle for two pours with a horizontal joint and after consulting Martin this was agreed with the proviso that the first pour was to be allowed to set and all laitance (loose concrete near the top of an underwater pour) was to be broken out be divers before completing the pier with the second pour. The result was satisfactory, but the time taken to get the pier cast was such that construction of one of the pier foundations for the Burra bridge had to be deferred till the next spring.

Supervision of these operations took up all my time in Shetland and from them I learned more about marine works to add to my Carradale experience. My stay was also notable for the diverse group I met in the Kveldsro Hotel which had been opened by Jim Williamson and his wife Irene the previous year. Jim was a very convivial host, especially in the bar of an evening and much fun was had with him and the other guests. His bar boasted one of the few taps of draft beer in Shetland at the time, the locals previously preferring bottled beer to drink with their whisky. But Jim had developed a taste for draft while on his national service with the RAF in England and on acquiring the hotel immediately installed draft beer of which he generally drank two pints to every one of his customers. The guests I particularly remember were a locum surgeon from somewhere in Devon who was in Shetland while the resident surgeon in the Gilbert Bain Hospital was on leave, a recently divorced lady lawyer from Glasgow involved in some project with the council and the newly appointed vicar of the Anglian church near the Kveldsro, of homosexual predilection, who was staying in the hotel while his house to be refurbished. There were others and with the local regulars as well a jolly time was had in the bar most evenings with Jim orchestrating proceedings.

By the end of October, the bridge contract was running down for the winter. Martin Broadgate was back to fitness and able to resume his full RE role. My short time in Shetland came to an end therefore and I travelled to Glasgow, starting my new job with Fairhurst as soon as I got there.

Ballachulish Bridge

My job title with Fairhurst was that of "Senior Engineer", an appellation I always thought something of a misnomer as senior to me meant someone older and more experienced than others which at 26 going on 27 I certainly was not. However, the term was used universally by firms of consulting engineers for those who had become chartered civil or structural engineers to distinguish them from others still to become qualified who were termed "Graduate Engineers". In late 1970 I would say roughly half the engineers working in Fairhurst's Glasgow head office were classed as senior and the other half graduate.

At the time I joined them Fairhurst were considered the pre-eminent bridge consultants in Scotland (by themselves at any rate) and certainly they could claim responsibility for an impressive range of bridge projects, such as the recently constructed mile-long bridge over the Tay at Dundee. Until around 15 years earlier the firm had been called F A MacDonald & Partners until the eponymous William Arthur Fairhurst, a blunt Yorkshireman, became senior partner and gained enough equity in the business to change the name to W A Fairhurst & Partners (or WAF & P as we knew it). He was a dedicated engineer who had published a respected book on the analysis of arch bridges which, in the days before computers supplanted the slide rule and all structural calculations were performed by hand, had become something of a bible for bridge engineers. In a story of him it was said that on his honeymoon he divided his time between preparation of the book and competing in some chess tournament in the place where the newlywed couple were staying. This may be apocryphal, but it probably captures the nature of the man quite well.

The big job in Fairhurst's design office in 1970 was the Charing Cross section of the Glasgow Inner Ring Road which included the prestressed concrete balanced cantilever Kingston Bridge over the River Clyde. This project was run by a partner called Gordon Farquhar. However, I was to be under another partner, Jimmy Brown, who was not involved in the Kingston Bridge or the Ring Road and instead, amongst other jobs was responsible for the design of the new bridge in Argyll to be built across the tidal narrows at Ballachulish. I was to work on this project, both on its design and construction, for virtually all the four years I was with Fairhurst.

The concept for the bridge had already been decided and some preliminary work had been done before I started with the firm. I joined a small team to take the project forward working under Ted Hinton as team leader reporting to Jimmy Brown, who I was to get to know quite well. Ted was easy to get on with and a very good engineer who went on to become a partner himself while I was with the organisation. The bridge was to be a through arch with a span across the Narrows of 600 feet or 183m as Britain was now moving to the SI system at that time and metric dimensions were beginning to be used. The south abutment was on a rocky outcrop and there was a long approach structure with a series of spans over a cobble beach on the north side. The main arch span across the water was to comprise two steel boxes with a composite steel and concrete road deck suspended on cables anchored to the arch ribs. My job was to design the steel box section ribs, it being considered by Fairhurst that I was qualified for this task by virtue of my involvement in steelwork on the bridges with Blyth & Blyth. The structural actions in these smaller bridges were completely different to those of the Ballachulish Bridge, but fortunately I had Ted Hinton to guide me and get me started.

The photomontages which Fairhurst had put together (no clever digitised mock-ups in those days) to present their ideas to the Scottish Development Department (the client) showed a bridge span of considerable aesthetic appeal in the dramatic setting of the Narrows with Glencoe in the background. I soon realised, however, as did all those who worked up the design, that the arch concept was flawed in one respect; there was no rock close to the ground surface on the north side to resist the large horizontal thrusts which would arise from arch action in the ribs. The south side on the rock outcrop was no problem, but on the north side there were deep deposits of gravels and sands with rock around 15m or more below ground level. While the gravels and sands comprised a dense stratum, it could not resist a horizontal thrust of over one thousand tonnes and a massive abutment was needed with a lot of raking piles driven to the underlying rock to counter the thrust. But this was not of immediate concern to me as my job lay with the steel box section ribs. These were predominately under compressive loading with some local bending as traffic loads moved across the deck below and the design of steel compression members and box girders was new to me. In fact, it was new to many others at that time as demonstrated by failures with loss of life of box girder bridges under construction at West Gate in Melbourne over the River Yarra and the Cleddau Bridge at Milford Haven in Wales, both of which resulted in searching public inquiries. There were also box girder failures in Germany and Austria. It was clear that a greater understanding of thin plate box girder behaviour was needed.

In the early 1970s design and construction of all steel bridges in the UK was governed by *BS 153 - Specification for Steel Girder Bridges*, issued by the British Standards Institution and augmented by some advice notes from the Department of Transport. In its four parts it contained general rules for materials and workmanship, live loading to be used on road and rail bridges, allowable stresses and basic specifications for steel bridge construction. Like its American equivalent AASHTO, it was an excellent code written in authoritative, easily understood terms. However, it did not cover all the requirements for large steel box girders assembled from thin plating, particularly load bearing diaphragms at the supports and interaction of all the parts of the boxes under bending and shear stresses. Following the failures, it became clear that action was needed, partly to preserve the international reputation of British engineers and partly to allay public concern. As a stop-gap measure, the outside lanes of all box girder bridges in the UK were closed to traffic until proper investigation of box girder behaviour could reveal the issues and remedies.

An independent committee was set up, under the chairmanship of Dr Merrison, Vice Chancellor of Bristol University, supported by a panel of specialists. Merrison was a physicist, not an engineer, and many thought him a strange choice, but he may have been an inspired appointment as he had no axe to grind and was free to pull together advice from the engineering specialists on the committee. Its main terms of reference were to consider whether methods of design and construction of major box girder bridges were satisfactory, to draw up interim technical guidance for bridge engineers for design and construction of such bridges and to recommend what further research and development into box girder bridges should be undertaken. Some interim design rules were issued in June 1971 and in October 1973 from the work of the committee and with results of much research conducted in parallel, the Department of Transport issued the main recommendations termed *Interim Design and Workmanship Rules for Steel Box Girder Bridges*. The new rules, while undoubtedly technically sound, seemed obscure and difficult to apply. However, they continued to be used until 1982, overlapping with the introduction of BS 5400 in 1978, a comprehensive code of practice covering all types of bridges and an excellent document which I used subsequently for many years until it was phased out prematurely for introduction of Eurocodes which lacked the clear focus of BS 5400.

During this period of flux in bridge design we were proceeding to work up and finalise the details of Ballachulish Bridge with its two large box section arch ribs which therefore fell into the category of bridges for which the Interim Rules had to be applied. In many ways our boxes could have continued to be governed by the simple rules of BS 153 as the structural actions in them were themselves simple being mainly the direct compressive loading of arch action. For example, BS 153 had a basic requirement to prevent local buckling failure of plating in compression which was for such plates to be restrained by stiffeners running in the direction of the compression stresses not more than 40 times the plate thickness apart. We had detailed the plating to keep the stiffener spacing well below this limit and we therefore had no concerns on local buckling failures. Nevertheless, we had to demonstrate that the plating would comply with the Interim Rules which ran to several pages of calculation in the case of local buckling. What we did, therefore in this and other parts of the design, was to obtain member sizes using BS 153 in the first instance and then to check what we had derived for compliance with the Interim Rules. All the design was done by hand calculation using the slide rule to obtain numerical values which by long practice we were able to do with great accuracy. This may seem archaic to today's generation of engineers, used to carrying out even the simplest of calculations by computer, but many major engineering structures were produced by similar methods which can stand comparison with the computer-designed structures of today. And of course, in a related field of engineering, the Spitfire was designed with the help of the slide rule as were the Wellington and Lancaster bombers as well as many other outstanding aircraft. Once we had worked through all the details there eventually emerged a finished design of the bridge ready to go out to tender by bridge builders of which the UK could still boast five major companies at that time, sadly most of them now no more.

During my early days in Fairhurst back in Glasgow, an event of great personal significance occurred in my life—I met Margo who was to become my wife. This was in January 1971 and in September of that year we became engaged and were married in August 1972 leading to 45 years of wonderfully happy marriage. As the purpose of this memoir is to set down something of my working life and travels in connection with it and not to write an autobiography, I would only say here that Margo was with me every step of the way from the time we met to when I retired and without her, none of it would have been worth a bean.

The five bidders all returned tenders for the arch bridge design, but one of them, my old friends from Shetland, CBE, offered an alternative design in addition at a 10 percent saving on the conforming design. Both CBE's conforming design and their alternative were cheaper than any of the other bids. This made their bid somewhat easier to assess. However, the alternative was only a tender design, not developed in the same detail as ours and CBE were not offering to construct it for a lump sum. In their offer it was to be adopted by the Engineer and to be subject to remeasurement. This posed a financial risk to the Employer as it had not been developed in the same detail as the conforming arch design.

CBE's alternative was a completely different type of bridge to ours. It comprised parallel steel trusses with the deck located between them, and it was continuous over three spans. It had a short side span on the south side of the crossing to the main pier on that side and a slightly longer side span from the north side main pier to an abutment. The main span of the trusses across the Narrows was 183 metres, the same as the span of the arch. Whereas our deck had been supported below the arch ribs on steel wire rope hangers, the cross beams of their deck were fixed on either side to the bottom chords of the trusses with longitudinal beams for the roadway deck spanning between the cross beams to form what is called a through-truss bridge. The chords of the trusses were relatively small steel boxes, 1.2m deep and 0.8m wide with plating of sufficient thickness to comply with the new design rules without the need for longitudinal stiffening. Although there was some internal welding for diaphragms needed to maintain the box in shape, the absence of stiffeners made fabrication easier and cheaper and had been developed from another successful tender by CBE for major strengthening of Britannia Bridge over the Menai Strait between Anglesey and the Welsh mainland.

After assessing the merits of the two options we concluded there was little to choose between them in terms of price as full development of the alternative would most likely involve additional steelwork, reducing the cost saving and bringing its cost close to that of the conforming design. The alternative had one big advantage in that it did away with the high horizontal forces on the north abutment, but to my eye it was undoubtedly less visually attractive than the arch, particularly as the depth between the chords was not constant and increased in depth at the main piers for structural reasons.

In view of the dramatic setting of the crossing the Scottish Development Department client had required the original arch design to be submitted to the Royal Fine Arts Commission from whom it had received a favourable response. It was therefore decided that CBE's truss design should also be put to the Commission. CBE prepared a photomontage of the truss which was submitted along with an updated photomontage of the arch. I think most people in Fairhurst thought the RFAC would come down on the side of the arch, but to our surprise the RFAC went for the truss in a unanimous decision (incidentally, Fairhurst himself was a member of the commission). While RFAC decisions were not legally binding, no-one in the SDD would ever dream of going against their view (this was in the happy days long before devolution in Scotland and the SDD was a department of the Scottish Office reporting to the Scottish secretary in Westminster). The truss it would be therefore, and the arch design was to be binned. We moved to some rather half-hearted discussion with CBE who were now in a strong position and contract details were soon finalised with the truss formally adopted as if it were the Engineer's design under the contract. Things had worked out splendidly for CBE; they had won the construction contract with their alternative which was to be subject to remeasurement, and the ultimate responsibility if things went wrong would lie with the Engineer. Fortunately, things did not go wrong, at least not seriously, but later I felt Fairhurst's partners were naïve to put the firm in a position to run such risks, or so I have come to think as I have grown older and wiser in managing construction contracts.

At this point another event took place which had a major effect on my working life. Ted Hinton was elevated to a partnership in Fairhurst, and I was moved up to replace him as the Team Leader for the bridge. I was to be responsible for all aspects of the final design in conjunction with CBE who would carry out the detailed calculations to be vetted and approved by my team. My main assistant was to be an up-and-coming engineer from Aberdeen called David Wright who had recently completed a period as RE on a steel bridge over the M80 which he had largely designed on his own under another Fairhurst partner. David was an excellent engineer, one of the best I ever met, with a quick and incisive mind, but he had alienated many people in the organisation by his manner. He did not take fools gladly and was not slow to point out where he thought others were going wrong. He played rugby at No. 8 for Clarkston's first XV and it was said of him in the club that he could start a fight in an empty room. However, we gradually warmed to each another and before long became very good friends.

At that time CBE had a strong design department attached to their works in Darlington. It was headed by a very competent engineer called Ian Dixon who was assisted by Mike Pargeter for a lot of the detailed work. They reported to a director called John Fletcher, a forceful character who had some family connection to CBE. A stream of calculations and design sketches started to flow from Darlington to Glasgow, gradually increasing in volume to a flood as Dixon and Pargeter got to grips with the detailed design. I had regular design meetings with them at their works in Darlington often accompanied by Jimmy Brown, after which John Fletcher would take us out to dinner in some country pub nearby. The design work was done at breakneck speed, as befits a contractor seeking to maximise his profits and Fletcher's driving to and from the pubs in the evening was similarly breakneck. I was stimulated by this and thoroughly enjoyed the process.

Then, one day David Wright came to me and said he thought there was a massive flaw in CBE's design concept which we had all missed. Being David the revelation of this had come to him, or so he said, whilst seated on the toilet where his best thinking was always done.

Truss bridges like CBE's design for Ballachulish comprise two distinct parts; the two parallel trusses and the deck which sits between their bottom chords. In CBE's design the deck system was made up of longitudinal beams integral with cross beams supported by the lower chords of the trusses on either side with bolted connections between deck cross beams and the chords. The cross beams and these connections were 12m apart along the lower chords. There was no flexibility in the connections, and this was the flaw picked up by David. Without any flexibility which would enable the deck system to move independently of the trusses, the deck would effectively become part of the bottom chords of the truss and participate in the total forces in these chords. There was no way of introducing flexibility into the bolted connections at the trusses and neither could they be designed to transmit the forces between the deck and the trusses for compatibility of the two parts. These compatibility forces would become very large indeed near the centre of the 183m main span of the bridge; the connections would simply be torn apart. That he was correct became apparent to me the more I thought about it. I quickly broke the news to Jimmy Brown and then to CBE. They were sceptical at first, but soon realised their design concept was in trouble.

Working with CBE we looked at possible ways around the problem. At first, we thought we might stiffen up the deck system into a horizontal truss of its own with almost continuous connection to the chords, but crude calculation showed this would need a large amount of extra steelwork which would cost a great deal. Whatever we looked at had serious cost implications and I began to think ironically that if the problem been apparent to us in tender evaluation, we would never have binned the arch design. However, a simple and quite elegant solution was found after much brainstorming between David and me on one hand and CBE's team on the other, which illustrates the power of coupling of several minds in problem solving.

The thing to do we realised was to raise the longitudinal deck beams to sit on movement bearings on the cross beams. The cross beams would still be rigidly connected to the trusses and would move with them as they responded to live loading on the deck, but the deck beams would not pick up these movements and the forces going with them as they would simply slide on the bearings. It would be necessary to anchor the deck beams to the trusses at one end of the bridge by means of fixed bearings so that longitudinal forces in the deck from all sources such as wind, vehicle braking and traction and temperature change would be resisted and transmitted on to the main bearings at the ends of the trusses, but that was a simple matter. An increase in the overall depth of the trusses between top and bottom chords by about one metre would also be necessary as the longitudinal beams would now sit over the cross beams rather than being integral with them and we had to maintain the necessary overhead clearance for vehicles on the deck. This involved an increase in costs as the truss verticals and diagonals became slightly longer, but it was a small price to pay to get the scheme out of jail.

There were no more serious issues with the design development which was completed rapidly from then on. Work on site in establishing the offices and site clearance had already started. On our side the Engineer's Representative or Resident Engineer was to be Jimmy Motion who had had several previous RE appointments with Fairhurst and he set about recruiting his staff. On CBE's side the Site Agent was to be Philip Harvey with whom I was involved with again later when I left Fairhurst. As the site staff were being recruited, I received a phone call from Martin Broadgate whose Shetland work had come to an end and was looking for a new appointment. I was happy to recommend him as assistant to Jimmy Motion and he was duly appointed as ARE.

As the design stage of Ballachulish Bridge had finally come to an end I was now responsible for head office management of the project. In practice this meant liaison with the site staff to clarify any design issues which could affect construction and to agree modifications as necessary to ensure the work on site proceeded as smoothly as possible. This was not a full-time job, and I was given another role as a checking engineer on bridge designs carried out by design teams both within Fairhurst and other consultants. While independent checking of bridge designs had always been undertaken within consulting engineer offices, as a result of the recent bridge failures the checking process had been formalised as a specific requirement for all bridge designs on behalf of the Department of Transport and the SDD. Three categories of checking had been identified: Category 1, simple structures which were to be checked by an engineer in the design group not directly involved in the part of original work being checked; Category 2, larger bridges which were broadly typical of their form of construction with no unusual features which were to be checked by a completely different in-house design team which had no involvement in the original design; and Category 3, major bridges or bridges with unusual design features which were to be checked by totally independent outside consultants. This system with some minor refinements remains in force in the UK to the present day and has been adopted around the world for bridges and extended to other civil engineering structures, particularly in ports and harbours and works with a high geotechnical content. In my subsequent career I have been involved in the design of many bridges and in independent checks of bridges designed by others both in the UK and overseas and I can say that, performed properly, it should provide assurance that proper design principals have been applied and the bridges are safe for public use. This has not always been the case as a recent tragic bridge failure in the United States demonstrated. Commercial pressures resulted in checking of a poorly designed bridge being skimped. A basic design flaw went undetected, and the structure collapsed under construction on to a freeway with considerable loss of life. In private sector works checking may also be carried out in a rather perfunctory manner. As a consultant after retirement from full-time employment I was involved recently as independent expert on a construction claim relating to the failure of a marine structure where the independent checking called for in the contract had not been implemented. While there had been no injury or loss of life associated with the failure, the resulting financial claim from the client was very costly for the consultant involved. Fortunately, things like that have not happened with highway and railway bridges in the UK where bridges are under the ultimate control of the transport authorities who always ensure the checking system is applied rigorously with formal certification signed off at each stage of the process.

Back in 1973 when Fairhurst formed its checking group of which I was the senior member, we were amongst the first to take on this role. In our case we were involved only with larger Category 2 and 3 structures for which the requirement was to carry out a truly independent check for each bridge passed to us. These might be steel and concrete composite structures, or ordinary reinforced concrete or prestressed concrete bridges with different types of support piers and other features. There were three of us in the group and I would split up the work between us and collate the results on completion when it would be typed up in a formal report submitted to the client, invariably the SDD, as the design checks we carried out were for trunk road bridges in Scotland for which the SDD was responsible. At the beginning we had no contact with the designers of the bridges and simply took their drawings and performed our own analysis to derive direct forces, bending moments and shear forces in the structure, then moved on to strength checks of all parts of the bridge. However, it was soon realised that dialogue between designer and checker was a good thing as it avoided misunderstandings as to the designer's intentions and the earlier dialogue was established the better for smooth completion of the checking process. The need to carry out a separate independent structural analysis was also relaxed. Checkers were permitted to use the results of the designer's analysis so long as the checker was satisfied that the method of analysis was appropriate for the type of bridge under consideration and the input data used in the analysis was relevant to the situation.

By now I had been with Fairhurst for over three years. Involvement in the construction of Ballachulish Bridge on the one hand, and experience of a wide range of bridges from the perspective of the checking group on the other, made me confident I could tackle the design of all but the largest cable-stayed or suspension bridge. I felt I was now ready to take on responsibility for whole projects with my own design teams under me, but as we moved into 1974 Fairhurst appointed two more partners and created some associates, and I was not among them. Discreet enquiries with Jimmy Brown indicated there would be no further promotions within the firm for some time. I became increasingly frustrated, particularly as the site work at Ballachulish was coming to an end and I was stuck in the bridge checking group, where the work to my mind was not as fulfilling as hands-on project design. Inevitably I started to cast my eyes around other consultants. At that point it did not occur to me to look beyond Glasgow as my wife and I were settled in our first house in Eaglesham. We had a young baby and my widowed mother in Langside was on her own and needed increasing help from us.

If I wanted to stay with bridges, there were only two other firms of any note in Scotland at the time: Babtie, Shaw & Morton and Crouch & Hogg. Babties were noted for dams and reservoirs and in terms of bridges they were the equal of Fairhurst, but they were a much bigger organisation overall with around 400 staff in various engineering disciplines. I had worked in their Water Department as a student one summer, and while I had nothing against them, I sensed promotion would be even slower than in Fairhurst. Crouch & Hogg (or "crouch in the bog" as we referred to them) were more noted for marine works than bridges, although they had an up-and-coming bridge department under a young partner called Alistair Wallace who always seemed to rub up Jimmy Brown the wrong way. This was perhaps because he was making a name for himself in the design of some notable bridges in Scotland. In fact, David Wright who had also become frustrated had recently jumped ship to go to Crouch & Hogg and work for Alistair Wallace. However, while I had become very friendly with David, I did not think there was room for the two of us in one organisation again and from what I knew of Crouch & Hogg, Alistair Wallace apart, they seemed like a smaller version of Babties in many ways.

Then a new name appeared in a job advert in the *Glasgow Herald*. A firm called Peter Fraenkel & Partners or PFP had opened an office in Glasgow and were looking for senior engineers with steelwork design experience in bridges or other heavy civil engineering structures. I applied and had an interview with the partner in charge of the Glasgow office, John Campbell. He told me the firm had been founded just two years before by Peter Fraenkel who had been one of the senior partners of the well-known London firm, Rendel Palmer & Tritton before resigning his partnership to set up on his own. John Campbell himself had been in RPT under Peter Fraenkel, but then had gone off to Canada to work for an engineering consultant there for some years before being head-hunted by Fraenkel to set up an office in Glasgow, partly to develop Scottish clients and partly to help overall expansion of the new business. John Campbell came over as a dynamic character with a get up and go attitude, and a slight transatlantic accent he had picked up from his time in Canada. The reason PFP were looking for people with heavy steelwork experience was a new job which had just been obtained in Tilbury Docks on the Thames involving a large floodgate at the entrance to the docks. He explained that, in addition to the Tilbury work, Fraenkel had already secured work in Scotland and had good prospects of obtaining overseas jobs in Africa and the Far East. I expressed some interest in all this and it was left that John Campbell would get in touch to take matters further when requirements for the floodgate had been more fully defined. That was in January 1974.

I heard no more for over six months and had assumed that, like many of these things, there had been a change of plan or someone else had been recruited. Then, out of the blue I got a telephone call from John Campbell. He was keen to discuss terms of an appointment for the specific role of designing the floodgate. I met him and he made me quite a good offer in terms of salary there and then, to be confirmed in writing in a day or two, which it was. Now I had to decide whether to take this up and leave the settled security of Fairhurst for a relatively unknown company.

While the salary was certainly better than what I was getting, there was no guarantee that promotion would come any more surely in the new firm, other than because it was new and perhaps there would be opportunities to grow with it. On this gut feeling, which was all it was, I decided to accept. In the event I probably made the right call, although things did go wrong much later. Eventually, the promotions which came were mostly from involvement in overseas projects. I joined the firm in late September 1974 and remained in it until I retired from full-time employment 43 years later.

In September 1974 when I started with PFP in the Glasgow office, there were around a dozen engineers there, one of whom, Ken Archibald joined the same day as me. Ken and I moved up through the firm's hierarchy in step and I came to know him very well socially in addition to working with him on many projects. There was also John Tainsh, who had started his career with Crouch & Hogg just prior to WW2 before leaving to join the army and then worked with John Coode & Partners, who were old established consulting engineers in London involved in major harbour projects throughout the world. I came to know John very well too as Engineer's Representative, in the UK and overseas in Hong Kong and Indonesia. Ken's schooling had been with George Herriot's in Edinburgh and John's with Glasgow Academicals and with mine at Glasgow High School there was a mutual interest in rugby. Finally, I met Willie Roxburgh who had just graduated from Glasgow with First Class Honours in Civil Engineering and as well as his academic achievements was a talented fly half with Kelvinside Academicals. While Ken and John sadly are no longer with us, Willie remains a friend to this day. I also met Peter Fraenkel for the first time. He happened to be in the office to discuss developments with John Campbell. He seemed a rather austere figure in a dark grey three-piece suit and after being introduced I was happy to get back to the desk which had been allocated to me.

I had been recruited by John Campbell specifically for the design of the Tilbury Docks Floodgate, based on what he considered to be my in-depth experience of heavy steelwork and on arrival I was handed the preliminary files on the project and started work. There were a few other jobs in the office, including a jetty for Cromarty Firth Port Authority who were expanding their facilities as a support base for North Sea oil exploration and drilling. Ken was to work on these projects while I was to focus entirely on Tilbury.

Tilbury Docks Floodgate and other UK projects

In dock and harbour engineering, floodgates, as their name implies have a primary function in protecting the docks against flood conditions while allowing passage into and out of the docks at other times. However, when the docks at Tilbury were constructed in the 1880s there was no need for protection against flooding. Certainly, the Thames could be prone to high water levels from flood surges emanating in the North Sea, but flood levels in the river were well known, and the quays in the docks were located safely above maximum flood level. The original entrance to the docks was at their eastern, downstream end where vessels could be accommodated in a tidal basin before entering the docks themselves. In the 1920s a new entrance was constructed at the western, upstream end with a lock and three sets of mitre gates so that vessels could enter and leave in all states of the tide and the old entrance fell into disuse and was closed in the early 1970s. At that point in time despite flood surges having become larger over the past 100 years, there was still no need for a floodgate. The highest surge levels could conceivably force the mitre gates open so that water would flow into the dock, but there was plenty capacity within the dock basin and no risk of overtopping.

What changed in 1974 was the impending construction of a flood barrier upstream of Tilbury Docks—the Thames Barrier as it became known. Bigger surges were now being experienced and these were putting large areas of east London at risk of flooding as well as much of the underground network. To safeguard London and the Underground, a barrier with rising sector floodgates was to be constructed across the river at Silvertown which could be raised on receipt of a flood warning to hold back the surge. The problem for Tilbury Docks was that on closure of this barrier the flood surge reaching it would be reflected down the river reinforcing the incoming surge. It was calculated that this could raise water levels by up to 3.5m above the historic highs for which the dock system had been designed. It was not just the docks on the river that would be at risk, but also the low-lying semi-urban areas and towns and villages along the banks of the Thames on both the Essex and Kent sides. Flood walls were therefore to be built along many miles of the lower Thames downstream of the Thames Barrier, augmented by secondary movable barriers at the docks and creeks to provide unbroken flood protection. The Tilbury Docks floodgate was to be the largest of these secondary barriers as the entrance to the docks was wider than any other dock entrance or creek.

At that time, the various London Docks were operated by the Port of London Authority which had responsibility for both maritime traffic and all facilities on the Thames. Each dock had a Dock Engineer reporting to the Director of the PLA with a sizeable staff under him to operate the facility and undertake routine maintenance. One of Peter Fraenkel's early achievements had been to recruit a former Director of the PLA, John Stanbury, as a partner and this resulted in a steady flow of work from the various Dock Engineers on the river. One of these jobs was the design of the floodgate for Tilbury Docks. In this case the ultimate client was the Anglian Water Authority (AWA) who had responsibility for flood control, but it had been agreed at government level that the PLA would take charge of floodgates where these were to be constructed on the dock estate.

The entrance lock at Tilbury Docks was built for the largest general cargo vessels using the docks in the 1920s and has three sets of mitre gates. In imperial dimensions the lock is 1,000 feet long, 110 feet wide with 45 feet minimum water depth over the mitre gate sills. It is one of the largest in the UK only rivalled by the entrance to the naval dockyard at Rosyth on the Forth which had been built to accommodate Dreadnought class battleships to counter the growing threat from the Imperial German Navy in the years leading up to the First World War. The floodgate would have to be able to close off this entrance effectively when a reflected surge was expected, but at the same time leave it completely clear the rest of the time.

Possibly the neatest way of doing this would have been by a hinged flap which would lie horizontally on the lock floor in normal conditions to be raised to a vertical position for a surge. But with continuous shipping movements through the lock day and night, major works to construct such a structure within the entrance would have effectively closed the docks and were therefore out of the question. Logically then the gate would have to be located on the side of the lock outside of the entrance altogether and somehow placed into the entrance when needed and removed again when the flood risk had passed. To complicate matters, the PLA had stipulated that nothing on the sides must come within 5m of the edge of the copes to be clear of bulbous bows of vessels going through the lock or be more than 5m high so that pilots and masters could see the lines of the copes ahead of them as they negotiated the lock entrance. Finally, the gate system, whatever it was, would have to be located outside the outermost mitre gates in the bell-mouth of the entrance where it was starting to widen, making the job of closure even more difficult.

This was what I was presented with on my arrival in the PFP office. There had been some ideas for a system mooted by various parties before I came, but as these were mostly of a Heath Robinson nature and completely impractical I had essentially a blank sheet to work on. I started to wonder what I had got myself into.

Somehow or other, working on my own, an idea started to gel, and I sketched out what I felt might be a workable solution which I showed to John Campbell. It would consist of a low-profile braced steel frame carrying under it a steel gate slung horizontally between the outer girders of the frame, hinged at one girder and held on cables at the other. The assembly would be parked clear of the lock entrance. In operation it would be moved up to the entrance on roller tracks which would be taken to 5m from the edge of the lock and the frame would then pass across the entrance in cantilever to the far side where it would pick up more roller tracks. Once the frame with the gate under it was straddling the lock, the gate would be lowered against four discrete sill blocks on the lock floor. These blocks would be located on the floor relative to the lines of roller tracks, so that the gate would be at an angle of just 3 degrees off the vertical when it was lowered against them. This would ensure a positive reaction force at the blocks to prevent the gate flapping back and forward under the action of waves in the lock entrance before the flood tide built up against it sufficiently to keep it closed by water pressure. Clearly there was a lot of working up to do, but John Campbell liked the idea, and in fact he became completely sold on it and that is almost exactly what was developed. We had to convince the PLA whose lock staff would be operating operate the floodgate, and to some extent the AWA, who were ultimately paying for the construction, but soon everyone bought into it. I proceeded to detailed design of the frame and gate almost entirely on my own with help on mechanical and electrical aspects from Geoff Nicklin the E&M partner in PFP who I became friendly with. Eric Phillips, another ex-RPT man who I worked with later in Bangkok, designed piled supports needed to support the frame along the roller tracks and on the sides of the lock, as the whole assembly weighed around 800t and ground conditions were poor. However, I can truly say the concept was mine alone as was the detailed design of the steelwork right down to all stiffeners and welding.

I recall one issue which caused me a great deal of difficulty before devising a workable solution. This was the fundamental incompatibility in stiffness between the steel plated gate structure when it was in a near vertical position and the two steel girders supporting it while they were spanning around 50m across the lock entrance. With the gate was in a horizontal position between the girders it would be relatively flexible and could accommodate deflections of the girders in cantilever and when spanning across the entrance. But when the gate was lowered it would effectively become almost infinitely stiff relatively to its supporting girders even though these were deep box beams. Without special measures of some sort, fixed hinge connections between the two would simply be torn apart. It was not just vertical deflection that was the problem as there would be relative movements along the line of the girders as well. I began to wonder if this was a showstopper for the concept. It was analogous to the incompatibility between the longitudinally rigid bridge deck and the trusses in Ballachulish Bridge which we had solved by allowing free movement between the deck and the cross beams of the trusses by sliding bridge bearings, and it was from the manner of solving that problem that a solution came to me. In the Tilbury floodgate system, it was necessary to spread the loading from the gate to several points along the support or hinge girder as I had called it and there were to be five hinges to provide this support and allow the gate to be rotated down into the operational position. To resolve the incompatibility which would then arise between the hinges, I converted them to floating hinges in a linkage arrangement and introduced thrust bearings capable of transmitting horizontal loading. With these linkages, free movement between the components would be assured as at Ballachulish. In the event the arrangement worked perfectly.

The contract for construction of the floodgate was put out to tender and was awarded to Butterley Engineering who had a long history of constructing dock gates and other heavy engineering works going back to the early days of canal and railway building in the UK, with Cleveland Bridge as subcontractor for the gate steelwork. This was the third time CBE had been a contractor on a job I was involved in and as at Ballachulish, Philip Harvey was CBE's site agent. Our ER on site was George Wright who moved on when the job finished to the permanent team looking after the main Thames Barrier. On site also, particularly during launching and lowering trials when there were some difficulties to be ironed out, was Robin Theobald who was still with PFP after many years of sterling service when I retired myself.

The gate was commissioned in 1980 and it performed well for forty years with just minor repairs from time to time. It was tested with a full launch once a month and used for real more and more frequently as sea levels rose. I suppose to some extent it made my name with the partners in Fraenkel right at the start of my career in the firm and earned me a reputation for making things work.

Despite the success of the design, there is one thing I would change if I could wind the clock back to when I was designing the floodgate in the 1970s. That is arrangement of the front rollers on the launching side of the entrance. These are in two banks of five under the lines of the main girders. As the frame with the gate moves across the entrance of the lock during the launching process in a progressively longer cantilever, more and more of the weight of the assembly passes through these rollers and especially the very last of the five rollers closest to the edge of the lock. Recognising this would happen I tried to build-in resilience in the supports by mounting the roller assemblies on natural rubber pads of moderate hardness which I expected would have sufficient "give" to equalise the reaction across all five of the rollers in each bank. This was only partially successful and the loading on the banks of rollers was by no means equalised with the end rollers still much more heavily loaded than the others. Throughout the forty years of operation the pads under these rollers have had to be replaced every few years as they are constantly degraded by repeated excessive deformation of the rubber. But of more concern, as the design was finalised, I became worried that the internal stiffeners inside the girders which distributed the roller reaction into the girders could fail if the reaction at the end rollers exceeded the buckling capacity of the stiffeners. There was no danger of the frame and gate tipping over as I had provided 60t of kentledge at the rear of the frame to hold it down with a positive restoring force until the front of the girders picked up the rollers on the far side of the entrance. Nevertheless, failure of the stiffeners would almost certainly have triggered buckling of the surrounding steel plating and resulted in collapse of the girders. After some last-minute calculations, I beefed up the stiffeners in the areas at risk as much as possible and managed to keep the stresses in them at a safe level. These changes to the framing caused some irritation and friction with Philip Harvey, but CBE carried them out as instructed and all was well in the end. Too late I realised that all this worry could have been avoided if the front roller banks had been mounted on trunnions. These would have rotated as necessary to match the slope of the girders in maximum cantilever and thus equalised the loading in all five rollers, which is what I would do if starting again—the benefits of hindsight.

In my first few years with PFP, all the jobs I worked on like the Tilbury Docks Floodgate were UK projects. Our second son was born in August 1975 and with two young children only two years apart in age, although I travelled to London a lot, I was around most of the time to give a hand looking after them. However, things were soon to change.

When I finished the design of the Tilbury floodgate and it was being constructed, I got involved in a range of other projects. These included failed boom anchorages in Liverpool Landing Stage where Fraenkel had been called in by Mersey Dock & Harbour Company to report on the failure and work with the design consultant, Harris & Sutherland, to devise a system of remedial works. The landing stage had come adrift in a storm in January 1976 when shore pintles securing the booms to the concrete pontoons had failed. At first it was thought that waves against the outside of the pontoons had caused excessive forces in the booms, but in fact it was geometric incompatibility between the series of booms that caused the failures. As the linked pontoons of the landing stage rose and fell with the waves running up the Mersey the ends of the booms at the pontoons all moved through different arcs and forces were set up in them which were well beyond the capacity of the shore supports to resist. These had failed one after the other as if becoming unzipped and the landing stage had started to drift off down the river on the tide, only to be stopped by the combined efforts of several port authority tugs. The answer was to build in resilience in the form of a rubber bearing and tension spring system at each of the shore supports of the booms, carefully designed to avoid the possibility of resonance and amplification of movements. The landing stage with this modification performed perfectly for many years until replaced recently with a new larger floating berthing structure.

The North Sea oil exploration which was starting when I was in Shetland with Blyth & Blyth had revealed large oil reserves under both British and Norwegian waters with a particularly sweet (low sulphur content), light crude oil ("Brent Crude") in the British sector. This was to become (and still is) the leading global price benchmark for crude oil and sets the price for a large proportion of internationally traded oil supplies. To get this desirable oil to shore pipelines were constructed from the Brent and Forties oil fields to Shetland where an oil terminal was built. Fraenkel achieved a major coup in obtaining the commission to design the terminal. It had three oil jetties and one gas jetty for export of product from the terminal as well as a tug harbour and ancillary facilities. Both the design and the construction of the jetties had been accomplished in only three years to meet the deadline for oil delivery through the new pipelines. John Tainsh was Engineer's Representative and I had been instrumental in fixing up Martin Broadgate as his deputy when his role on Ballachulish Bridge with Fairhurst had come to an end. I was not involved in the design of the jetties which was carried out in London under Roger Postlethwaite, the youngest and most dynamic of the Fraenkel partners, although I had a great deal of involvement on the jetties much later in my career. The only thing I did at that time was to design steel fender frames on the head of the gas jetty to cater for small gas tankers which were too short to use the main berthing dolphins.

I then designed a slipway and boat shelter for the ferry *MV Good Shepherd* which served Fair Isle on a twice weekly service operating out of Sumburgh in Shetland 15 miles to the north. At the time there was only a rudimentary berth for the ferry on Fair Isle in a bay called North Haven. This bay was untenable in easterly or northerly gales. Later, we provided a breakwater at the entrance to North Haven, but at that time the council had insufficient money to pay for constructing a breakwater and a stopgap arrangement for berthing the ferry in North Haven was needed.

The solution which was adopted was I think rather neat, and I believe the idea came from John Tainsh. This was to cut a slot for the ferry in a rock cliff behind the beach and to locate a slipway directly in front of the slot with the slipway rails extending across the beach and into the slot. A landing stage was provided alongside the slipway for the vessel to berth against in normal operation. If the weather deteriorated, however, the vessel could be slipped on a steel cradle fitted with flanged rollers and quickly winched up the beach and into the protection of the slot. My slipway design incorporated precast concrete beams running down into the water to sufficient depth at low tide to slip the ferry with the cradle running on steel bullhead rails fixed to the beams. Foundations for the beams were in the form of concrete manhole ring plinths taken down to rock and filled with concrete for stability. This arrangement worked well and was eventually augmented in the 1990s with a breakwater across the entrance to the bay.

Then my role in the firm changed radically.

Getting the Knees Brown

We travel not for trafficking alone,
. By hotter winds our fiery hearts are fanned:
For lust of knowing what should not be known
We take the Golden Road to Samarkand.
From Golden Road to Samarkand, James Elroy Flecker

The change came as we moved into 1977. Peter Fraenkel had made some rather astute appointments in other partners who had good connections with potential clients, particularly overseas. One of these partners was John Stanbury who had already played a big part in obtaining the floodgate work in London, and he also had many contacts from his days in the Ministry of Works looking after engineering requirements for British military establishments overseas. For some reason, these included admirals in the Thai Navy, even though Thailand had never been part of the British empire. From these contacts a stream of new work emerged, in the Far East and West Africa. So much so that by 1980, 80% of the firm's jobs were overseas. I was to become heavily involved in much of this work.

It started for me with designing a steel flyover at Klong Ton in Bangkok. This was over a busy junction and a klong in a street off Sukhumvit Road, one the main traffic arteries in the city. I spent an intensive six weeks in Bangkok doing this job, working in an office of a local firm, Metropolitan Engineering Consultants, who provided me with a team of draughtsmen to produce working drawings of the flyover, but with no engineering help other than a surveyor to fix coordinates of the flyover centre line and obtain levels for clearance over the junction. I stayed in a flat rented by PFP in Soi Lang Suan with Peter French who was stationed in Bangkok at the time. Peter reported to John Stanbury and chased up opportunities for the firm (such as Klong Ton Flyover). He had been in Bangkok for about nine months and knew his way around and thus was a good guide to the delights of the city at that time. This was before Bangkok became flooded with Western and Australian tourists and it still had the authentic, exotic feel of the East. Not least, I developed a taste for Thai food which has remained with me to the present day. A good introduction to a part of the world I was to get to know well.

The flyover was supposed to be demountable, although as far as I know it is still there to this day. I had a free hand with its design and opted for mostly steel throughout the structure which, being largely bolted, would be easier to dismantle than reinforced concrete. My design consisted of a steel deck of universal beams and profiled galvanised steel sheeting with concrete infill, steel crosshead beams and single large diameter steel tubular columns to provide minimum obstruction to traffic on the junction below. All spans were designed to be simply supported with expansion joints between them. This would not give the best riding quality for vehicles on the flyover, but it would simplify matters if the structure were ever to be dismantled. The modular form of construction presented few design problems and with good draughtsmen I managed to complete the assignment within the six weeks allocated.

Biu to Maiduguri Road

Almost immediately on my return to the UK I was off again, this time to Kano in Northern Nigeria to design bridges on the Biu to Maiduguri Road, which was a World Bank financed project. This was also an intensive assignment. During a stay of just under three months in our office in Kano, I surveyed six river crossings and carried out conceptual and preliminary design of the bridges. In fact, I designed twelve bridges on the project as contractors bidding for the work were to be offered the option of either steel composite decks or precast prestressed concrete decks for each crossing.

Arriving in Kano late at night I had my first experience of the pervasive "dash" without which nothing could be done in Nigeria. For some inexplicable reason, our luggage had been offloaded from the aircraft some distance from the terminal building and in order to retrieve one's cases, each of the passengers had to pay the head of the ground staff dash of 10 Naira. I then had some difficulty getting a taxi (nobody met me) and finding the place I was to stay, in dark, poorly lit streets. This was a bungalow which housed several road engineers working on the project. These were Colin Campbell, Miroslav Huq and David Daniels. In the team in Kano there were also some engineers who were there on an accompanied basis: Roger Base who was running the show, Louis Akroyd a soils specialist, John Smith a senior highway engineer and Mike Butterfield, also a highway engineer whose wife had brought her cats out with her. Everyone became members of the Kano Club which included a golf course with oiled sand greens (or "browns"), a swimming pool of doubtful water quality and a roofless squash court constructed in concrete, as well as the usual facilities of former colonial clubs. Playing squash around noon in the open concrete court with the sun beating down directly overhead was a unique experience. The golf course had a bar attached to it (naturally the Golf Bar) presided over by Moses the head barman with several boys under him for busy times, which were most nights, with some of the most regular patrons well-to-do local Muslims who had been on the hadj. I spent quite a bit of time there, sometimes after playing squash when urgently in need of rehydration, sometimes after a round of golf with borrowed clubs and on many other occasions. We drank Star Lager, often in company of some of the hadjis, which was supplied in large bottles containing around a pint and a third of beer. It was a strong brew, and one had to be careful driving back to the bungalow on dimly lit, poorly maintained roads. Driving was made even more fraught by the fact that, for some reason never fully explained, the Nigerian government had decided a year or so before to switch from driving on the left to driving on the right. Some residents had clearly not got their heads around this as evidenced by the wrecked vehicles littering the road verges throughout Kano. Of course, the road layouts were not designed for driving on the right, which compounded the problems. Outside of the towns, driving was especially dangerous. Most of the roads were single lane, the so-called "narrow tar" where the only rule of the road when a vehicle met another coming from the opposite direction, was for the smaller vehicle to give way and get off the road. Confusion as to who should give way to whom resulted in many fatal crashes. Driving in the hours of darkness was to be avoided at all costs.

In the main, however, I found Kano and the north of Nigeria quite a pleasant place to stay. Kano had its share of beggars with all manner of afflictions, but although summer temperatures rose to over 40 degrees Centigrade, the humidity was low, and the north was relatively unpolluted unlike Lagos in the south. Kano was a Muslim city as was the north in general. However, there were still quite a lot of

Christians living there at that time, before they were driven out by the introduction of Sharia law and repeated anti-Christian riots and killing twenty years later, and it was still largely tolerant and peaceful in 1977. Theoretically, infidels (unbelievers) like us, were not supposed to remain within the city walls after sunset, but as many facilities were located inside the walls this requirement was not enforced. Kano also had several restaurants, mostly Lebanese-owned, and one in particular, "Le Cirque Bleu" was surprisingly good, giving me a taste for Lebanese food.

The road I was involved in was in Borno state about 400km east of Kano. It was to run from an administrative town called Biu in the south of the state to the capital Maiduguri, 170km to the north and near the border with Chad. Biu is on a plateau at an elevation of over 600m and Maiduguri is about 300m lower, and the route gradually descended as it ran northwards. Even then Maiduguri was a sizeable place, and it now has grown into a major city with a population over one million, sadly beset by violence instigated by the jihadist terrorist organisation, *Boko Haram*. Nevertheless, it was and presumably still is, an appealing city with the broad streets laid out by the British colonial administration and planted with neem trees along the road verges and footpaths which provide shade from the fierce summer sun.

About a month of the early part of my stay in Nigeria was spent under canvass in a camp in the bush where it was my job to identify the best locations for river crossings and survey them in sufficient detail for preliminary design of the bridges. This was a key role as the locations of the crossings essentially defined the alignment of the road. The bush along the route was typical of the featureless, semi-arid Sahel region, lying between the Sahara Desert to the north and savannah grasslands to the south. Nomadic Fulani herdsmen were a common sight in the bush appearing in their family groups, driving cattle south from the Sahel in search of more abundant grass for grazing and moving back north again after the rains. There was a lot of tree cover in the form of Acacia and Baobab, and it was easy to get lost as there were few topographical features for reference. We relied largely on bush tracks to get around which disappeared in some places forcing the vehicles to thread their way through the bush. Existing mapping was basic, and we moved on compass bearings to identify the river crossings. It was not always easy to find the rivers as they were seasonal and near the end of the long dry season when I was surveying them, they were mostly dried up except for occasional muddy pools.

There was only one town of note, Damboa, which was midway between Biu and Maiduguri, although there were many small settlements of traditional beehive-shaped mud-walled dwellings. A little cotton was grown for export, with subsistence agriculture and slash-and-burn cultivation generally and the better off inhabitants kept scrawny long-horn cattle which roamed the bush near the settlements. We set up the camp a little to the north of Damboa with permission from the headman of a settlement nearby. From our camp we could cover the route to its ends north or south within a day and there was a ready supply of labour in the settlement for digging trial pits along the route to determine soil conditions for the road pavement design. We were well supplied with food which could be obtained in Maiduguri, and we also had plenty beer and soft drinks. The main problem was the fuel shortage in Nigeria at that time as there was only one refinery in the country which was often out of action and our drivers could sometimes spend days in queues in Maiduguri waiting for fuel to come in so they could fill up jerry cans and drums with petrol and diesel for our vehicles. These were Series II Land Rovers and long and short wheelbase Toyota Land Cruisers.

My vehicle was a Land Rover for which I had a driver called Ibrahim, a tall friendly Nigerian who was with me throughout and we often shared the driving which could be very tiring in the bush. At that time Land Rovers had 2.25 litre petrol engines which were mechanically sound, but somewhat underpowered as compared with the 4.2 litre diesel engines in the Land Cruisers. This could sometimes give problems in getting the vehicle up steep banks of the dry rivers which the Land Cruisers could accomplish with ease. The other issue which affected both types of vehicles was frequent punctures from thorns in the bush and we always carried at least two spare wheels on any journey to bridge sites.

There were two English VSO (Voluntary Service Overseas) teachers stationed in Damboa who taught in the little school there. They were young women in their early twenties on what amounted to a gap year who lived in an old bungalow near the school. As most of the bridge sites lay south of Damboa on the rising ground towards Biu, I would often stop at their place for a cup of tea on the way back to the camp after a bridge survey and drop off some beer for them. On occasion, if surveying late in the afternoon, I could sleep overnight in a spare room in their bungalow while Ibrahim found somewhere to stay in Damboa with 20 Naira I would give him for this purpose. A posting like this for two young women sadly would be impossible today with *Boko Haram* around as they would be kidnapped and ransomed or worse in short order.

The first thing to do was to find the rivers, then identify straight reaches for crossings and finally survey the chosen crossing points to enable a preliminary bridge design to be carried out. It was important to be able to fix the position of these crossings by some feature on the ground which could be seen on aerial photos for later design of the road alignment. I would make an estimate of the storm discharge at each site, based on the approximate catchment area and the river bed gradient from what mapping we had, using a rainfall intensity of 150mm per hour. This approach was probably grossly conservative, although intense rainstorms did occur in the north during the wet season. Invariably this would show the river to run bank full and with Ibrahim's help I would seek out the headman in the nearest settlement for information on any flooding from overtopping. As the approach embankments to the bridge would constrict the flow if there was overtopping, I would add 0.6m to the bank level plus another 1.0m of freeboard to get a safe level for the underside of the structure. I used the same reasoning in East Malaysia on many bridges there and as far as I know none of my bridges in either country have suffered hydraulic problems in service.

By the time I had nearly finished the surveys it was early May, and the temperature was topping 40C most days. We had to drink copious amounts of water while out in these temperatures and carried a goatskin containing three gallons hanging at the tailgate of the Land Rover. The skin was slightly porous and evaporation through the skin was supposed to keep the water it contained reasonably cool. On one of my last days in the camp I had three bridge sites to visit to get final details for the design. These were all some distance south of Damboa. We set off early just after dawn while it was still reasonably cool, but by the time we reached the first site the sun was well up and the temperature was climbing rapidly. It took about an hour to get the information I needed with Ibrahim to help me and then we headed on to the next site. We were already becoming very thirsty and started to drink from the goatskin. Another hour was needed at the second site and then we headed for the third with the goatskin half empty. When we finished at the third site it was noon and the temperature must have been in the mid-forties. By now the sun was directly overhead and we could not go more than five minutes without drinking from the goatskin which we had nearly emptied by the time we started back for the camp. When we did get back and had put away the survey equipment, we both drank cup after cup of tea and later I must have put away six cans of beer sitting in the shade outside my tent. I had become so dehydrated on the morning trip that despite drinking one and a half gallons of water, innumerable cups of tea and a lot of beer, it was 10pm before I passed any water that day and even then, only a meagre dribble. However, I had got everything I needed for the office work in Kano.

On the evening before we broke camp, we packed up our equipment and stowed it in the vehicles ready for the journey back to Kano the following day and had a bit of a party to finish up our beer, to which we invited the headman of the settlement and some elders. Anything we no longer needed we left with the locals who had been most helpful throughout and despite the camp often being largely unattended during the day nothing had ever been stolen. We often joked about Nigerians that if they found something, they would "eat it, steal it or shag it, and not necessarily in that order". This may have been true of some of the inhabitants of the cities, but it could not be further from the truth of those in the bush who were completely honest. In my experience it is the same with rural peoples the world over.

Back in Kano, I got down to preliminary design of the bridges with the information I had obtained. This was not difficult. There were no big river crossings and the bridge spans ranged from 18m to 25m which would suit simply supported prestressed concrete beams or composite steel and concrete deck bridges. Abutments in either case would be clear of the banks of the rivers which could be protected against scour at the bridge sites by gabions. The bridges were designed to UK Department of Transport HA Loading and 45 units of HB loading, much as they would be in the UK.

The night before I flew back to the UK, I enjoyed a night with the road engineers at the Golf Bar in the Club. To go with the beer, I ate some of what we called "serai", spatchcocked goat meat, grilled and liberally spiced with cumin, which we would often order at the bar. I must have had a dodgy one because overnight and the next day I had severe diarrhoea and only just managed to catch the evening flight to London. In fact, the diarrhoea did not clear up as these things usually do and came back with a vengeance when I was in Hong Kong some weeks later. I was on the point of seeking proper medical advice in Hong Kong, but I thought I would see if I could get some help from a pharmacy near the office before I did so. The Chinese owner listened to my story and said the problem would not go away unless I took some antibiotics. He handed me a plain packet with instructions to take one of the pills it contained four times a day. They worked and the diarrhoea cleared up almost immediately. These were the days when you could get any antibiotics you wanted over the counter in Hong Kong, and I used that pharmacy for years to get antibiotics quickly for sore throats and so on.

Sha Tin to Tai Po Trunk Road

PFP had opened an office in Hong Kong in 1974 and had scratched around trying to drum up business, not very successfully, for three years until in 1977 a large job was finally obtained. In late 1976 I had been involved in a proposal for the design of a major bridge between Aberdeen on the south side of Hong Kong Island and Ap Lei Chau (duck's tongue island in Cantonese). This was in a joint venture between PFP and Tony Gee & Partners and Tony Gee himself had come up to the Glasgow office to work with me on the proposal. He had set up on his own after a successful career with Maunsell in London and had been making a name for himself as a bridge consultant. Between us we came up with a novel concept which involved constructing the bridge in two halves largely on land and then slewing them through 90 degrees to meet in the centre and so avoid the need for expensive work over water. As it happened, we were just pipped for the job by Maunsell who, along with Scott Wilson were the leading civil engineering consultants in Hong Kong in the 1970s. However, although we had not been successful, our proposal had impressed the PWD who let it be known that we would be in line for a commission if a suitable project came up.

What came up was the Sha Tin to Tai Po Road Trunk Road as it was called then, located in the New Territories, the area north of the Kowloon peninsula stretching up to the Chinese border. It was later known as the Tolo Highway as it ran alongside the sea inlet known as Tolo Harbour. The road would only be 10km long, but the terrain along the route made it the most ambitious and expensive highway project to be undertaken in Hong Kong up to that time. My job would be the design of the bridges, starting with the concepts in a feasibility study which commenced in late June 1977. I was very keen to go to Hong Kong, but I had made it clear to the partners I would only do so if my family could accompany me. I had left my wife alone with two small boys for nearly six months and although her parents were not too far away, she was really on her own. She had not complained and had just got on with it, but I knew it would be unfair to leave her for another extended period, neither could I contemplate a further lengthy separation myself. I felt I was in a strong position with this demand as I was the only experienced bridge engineer in the organisation and so it turned out. It was agreed I would go on my own almost right away and my wife and the boys would follow two weeks later, and I would get an additional per diem allowance to cover the family costs. This suited me as it would give time to arrange accommodation and get a feel for the place.

Towards the end of June 1977 therefore, I flew out to Hong Kong on a British Airways flight taking with me, as well as my own luggage, a large cardboard box containing proposals for consultancy services on a World Bank financed road project in Malaysia. It was an eventful flight to say the least. In those days there were no direct flights to the Far East, and we had four scheduled stops for refuelling. On the second of those, in Doha, a problem was discovered in one of the engines. Attempts to fix this failed and another engine had to be brought out from London. The passengers were offloaded, along with luggage, and taken to the Gulf Hotel near the airport. One day later with a new engine fitted we were about to re-board the aircraft when an aircraft in flight was hijacked by some terrorists and put down at Doha as it was running out of fuel. It was immediately surrounded by detachments of the Qatari army and the airport was shut down. Back to the Gulf Hotel we went, this time for a day and a half, until the hijackers surrendered after protracted negotiations and released the passengers on the aircraft. Eventually we got off on the next legs to Karachi and Colombo and finally on to Hong Kong, two and a half days late.

When I arrived at Kai Tak Airport it was two in the morning, and I took a taxi to the hotel which I had been booked into. It was in Causeway Bay in Hong Kong Island, and to get there we travelled through the new Harbour Tunnel. Given that our office was in Tsim Sha Tsui it would have made more sense to have put me in an hotel on Kowloon-side, but as I was not intending to stay in an hotel for long, I was not too bothered. My first impression of Hong Kong was how busy the streets were even at night. After the long and tiring flight from London I slept in till late the next morning and when I looked out again at around 11am the streets around the hotel were now jam-packed with pedestrians. This was a place that never slept.

And thus began, for me, a deep attachment to Hong Kong which developed as a result of my involvement there during the next eleven years and on subsequent visits and shorter assignments up to the time of my retirement forty years later. This attachment is not uncommon in *Gweilos* who lived and worked in Hong Kong, particularly those of us who were there under the British administration prior to the handover to China in 1997. It might be argued that our feeling for Hong Kong simply stems from the good times most of us had there with an active social life amongst like-minded people, with clubs and servants, a mostly pleasant climate and good earning potential in a low tax environment. While it is true that most of us did have a good time whilst there, perhaps enjoying the best days of our lives, I think the feeling runs deeper. It does with me at any rate.

It may be something to do with working in Hong Kong as a civil engineer. As a profession, we had a key role in shaping the infrastructure necessary for Hong Kong's development. Hong Kong was impoverished after the Japanese occupation, but with the restoration of British rule in 1945 quickly got back on its feet as labour and capital flooded into the colony and it filled up with immigrants and people with money from Shanghai after the defeat of the Chang Kai-shek's Kuomintang by the Communists in 1949. In the space of 50 years, it was transformed to the modern and thriving city territory which was handed back to China in 1997. With the Chinese in Hong Kong making money and paying tax to the colonial government to finance development, we engineered the infrastructure, building urban and rural roads and flyovers, stabilising steep slopes for massive public housing developments, building deep water quays for container terminals, providing ferry piers and breakwaters on remote outer islands, and constructing a new international airport along with many more civil engineering works. In doing so, perhaps we engineers came to identify with the place we worked in more than expatriates in other sectors of the economy such as banking and finance who came in increasing numbers as Hong Kong became an important financial centre and grew more affluent.

By 1977, when I first went there, Hong Kong was the epitome of unfettered capitalism. The title of Richard Hughes's 1976 book *Borrowed Place, Borrowed Time*, which as Hughes freely acknowledges was taken from a quotation by Han Suyin (author of the excellent book *A Many-Splendoured Thing*) some years earlier, aptly catches its unique situation. In his book Hughes describes Hong Kong as "an impudent rambunctious free-booting colony, naked and unashamed, devoid of self-pity, regrets or fear of the future". This sums it up in the 1970s, although as far as I could gather it had been even more rambunctious in the years immediately following 1949 when money from Shanghai had flooded in. The relationship between the British and the Hong Kong Chinese was complex, but for two races equally characterised by a high degree of xenophobia, it was a symbiotic one. The Chinese in Hong Kong thrived under the rule of law and the light-touch laissez-faire approach of the colonial government which left them to prosper in a way they could scarcely have done anywhere else. We, the British, benefited by the opportunities offered in the unique environment of Hong Kong powered by the industry of the Chinese. In his book Richard Hughes neatly encapsulates this in a short paragraph:

Hong Kong's success is the living evidence of what Chinese labour and talent, given free opportunity, can achieve in artificial conditions, under foreign administrators, and with a nineteenth-century economic system.

However, I very soon came to realise also that Hong Kong was first and foremost a Chinese city, not that I minded that in any way. Despite being in a British colony, the Chinese had not lost their certain belief that they were superior to the foreign devil Westerners in their midst. With over forty years of involvement with the Chinese, not only in Hong Kong, but elsewhere in the East, I am inclined to think they are in some respects, but not in others and I suspect this is the view of most *Gweilos* who have worked with them over any length of time.

(As a point of note, in referring to Hong Kong as a geographic entity I have used the word "colony" throughout, although even as far back as 1977 the British administration had decreed that the word was not to be used in communications on government projects and was to be substituted by "territory". I always felt the word territory inadequate to describe the unique British-run Chinese city of those days and officially anyway, Hong Kong remained a colony until 1997.)

The Trunk Road was to run from Sha Tin along the western shore of Tolo Harbour to Tai Po to the north. Tolo Harbour or *Tai Po Hoi* (Cantonese: Tai Po Sea) is a long sea inlet in the New Territories. Both Sha Tin and Tai Po had been designated "New Towns" by the Hong Kong government to house the colony's rapidly expanding population. Maunsell was responsible for developing infrastructure for Sha Tin and Scott Wilson for Tai Po and the Fraenkel project fitted in between the two. Some years later I learned the appointment of a new engineering consultant on this project was a deliberate strategy on the part of the PWD which was soon to be split up into several departments such as the New Territories Development Department (NTDD) who became our client. It seems they wanted a "third force" roads consultant in NT to avoid having too many eggs in the one or other of the Maunsell/Scott Wilson baskets and appointing PFP for the Trunk Road project neatly split up the established firms. Unfortunately, for various reasons, while the design and construction of the Trunk Road was a resounding success, PFP never really grasped the opportunity being offered. One of these reasons was the way the office was run.

A highway engineer with wide experience of roads in Africa and the UK was put in charge of the office and at the same time led the design of the project. He was a good engineer and well suited to the role of project director on a major government highways project, but he neglected to promote the firm with other potential clients from a position of strength as the consultant for the largest roads project undertaken in Hong Kong up to that time. Instead, he focused almost entirely on the design of the Trunk Road and no meaningful promotion was carried out while he was there. In fairness to him, the Trunk Road was a demanding project and the recently set up NTDD a demanding client. The main blame probably lay with the partners in London in not appointing a more dynamic manager to run the office with partner status in the firm, and in not providing proper funding for the office even if that would have meant running at a loss while other work was chased up. Instead, the books had to be balanced against the Trunk Road fee income and things were kept on a very tight rein. As a result, the opportunity to build a strong position in Hong Kong went begging.

The terrain between Sha Tin and Tai Po is extremely hilly and climbs from the western shoreline of Tolo Harbour to the mountains of Kowloon, culminating in Tai Mo Shan, translating rather prosaically in Cantonese as big flat mountain and at 957m the highest peak in Hong Kong. In fact, a small job PFP had done before landing the Trunk Road appointment had been to design an upgrade for the road to the summit of Tai Mo Shan where there was a radar station, built originally for the RAF and operated in 1977 by the Royal Observatory. The Kowloon-Canton railway which also ran between Sha Tin and Tai Po on its way to mainland China had been located on the flattest land available which was at the base of the slopes near the shore. The choice for the route of the new road, therefore, was along the steep side-long hillsides where there was an existing single carriageway road, or on reclaimed land in Tolo Harbour itself. Our initial studies showed the latter option likely to be the cheaper of the two and this was accepted by the NTDD.

Soon after I arrived, I managed to rent accommodation in a flat in a low-rise block in Silver Strand on the Clearwater Bay Road in the north-eastern New Territories, with a car thrown in for a nominal additional charge. My wife and the boys (4 and 2 at the time) arrived at the beginning of July to the heat and humidity of a Hong Kong summer afternoon. While Margo waited with John in the relative cool of the Kai Tak Airport terminal building, David accompanied me to collect the car which was parked in the open. He must have got a shock when he stepped out of the terminal as I remember sweat breaking out on him and pouring from his face and hair. He had never experienced real heat and humidity before, and it was a particularly hot summer day even by Hong Kong standards. Such is the resilience of young children, however, that both he and John had completely acclimatised to the conditions within hours of their arrival.

My role in the team which PFP assembled in Hong Kong for the feasibility study was concept design of the bridges. I was also involved in studies to determine the optimum level of the highway for protection against flooding in extreme typhoon induced surges in Tolo Harbour. This latter work was the responsibility of Hans Eilenberg, a German Jew who had come to Britain as a refugee from the Nazis in the 1930s and had fought with the British Army in the war, losing a leg in the conflict. He was a brilliant engineer who worked with us on a freelance basis who was to become a good friend.

There were twelve locations along the route where bridges were required and all but one of these were overbridges at interchanges or where minor roads had to cross the Trunk Road. The exception was a bridge carrying the Trunk Road over a river which we named Tai Hang Bridge (*Tai Hang* translating as big water in Cantonese). The other bridges were also given names after features local to the bridge, for example Banyan Bridge, located at the Tai Po end of the route. This bridge was named after the large Chinese Banyan tree in the garden of Island House, the residence of the secretary for the New Territories, at the time David Akers-Jones, later Sir David. He had played a large part in designating the villages of Sha Tin and Tai Po "new towns", initially to resettle squatter dwellers from the hillsides of Hong Kong and he took a special interest in our work in designing the new road.

At that time steel bridges were not favoured in Hong Kong on the grounds they would require frequent maintenance painting to prevent corrosion in the periods of rainy weather interspersed with hot sunshine prevailing over the summer months. Also, there were no rolling mills in Hong Kong and although increasing quantities of steel were being produced in China its quality was uncertain and, in any case, China was still riven by political arguments in the wake of the Mao Tse-tung Cultural Revolution (Chairman Mao had died only six months before in December 1976). Structural steel for bridges would therefore have had to be imported from Europe or Japan. On the other hand, there were large supplies of crushed granite aggregates for concrete. In view of this it seemed that all the bridges were best constructed in concrete in one form or another. Precast, prestressed concrete beam bridges were popular in the UK. They were the mainstay of bridge construction on the motorway programme at that time and were also beginning to be used in Hong Kong. In the UK, the bridge beams were mass produced in long-line pre-tensioning beds by several manufacturers and transported by road to the bridge sites, but as there were no such facilities in Hong Kong, bridge beams were generally cast in situ at each individual bridge site and were then post-tensioned. I thought we could improve on this with twelve bridges to construct if we maximised the use of a uniform section of precast prestressed concrete beam and produced all beams in a central casting yard to provide economies of scale. As the scheme evolved, several other bridges were added subsequently which were not suited to precast beams for one reason or another and these were of different forms of construction. However, most of the bridges on the Trunk Road had precast concrete beam decks with over 400 beams in total.

While there was abundant concrete aggregate available, the granite rock in Hong Kong tended to produce an angular aggregate when crushed resulting in a harsh concrete which did not flow into the forms as well as concrete made with river gravels commonly used in the UK. The standard motorway "M" beam in the UK had a vertical web only 160mm wide and relatively free flowing river gravel aggregate concrete could flow into these webs easily. This might not be the case with the harsher granite aggregate concrete, however, which could be held up in the webs by the prestressing ducts causing voids and porosity in the concrete after casting. To get around this I increased the thickness of the webs to 200mm and hoped this would be sufficient as there would be a weight penalty if the webs were made much thicker with little strength gain to offset it. Nevertheless, this remained a concern to me, and these worries resurfaced when the first beam was cast once the construction contract was underway.

A major decision to be made at the outset of the study was the optimum level for the road carriageway. This had to be above flood level in typhoon surges which could raise sea level several metres above regular high water. Records of typhoons and accompanying surges were made available to us by the staff of the Royal Observatory which was located just behind our office in Tsim Sha Tsui. From these Hans carried out a joint probability analysis of the return periods for typhoon severity, surge effects and high spring tides. I helped him arrange the records in a suitable form for his analysis and did some numerical checking of the calculations. I learned a lot about Hong Kong typhoons from the Observatory records which dated back to before WW2. The wind strength in a typhoon depends on the barometric pressure at its centre or "eye". The most severe ones as far as Hong Kong is concerned almost invariably pass close to the south-west of Hong Kong, bringing the strongest winds from south-east over open sea and pushing up the biggest surges. There was no question that typhoon surges were a real threat to the road as evidenced by the records of a massive unnamed typhoon in early September 1937 which had driven an enormous surge into Tolo Harbour drowning at least 10,000 people in the Sha Tin area. We concluded that a minimum carriageway level of

+6.1m above Principal Datum (PD) should be adopted to minimise risk of flooding from wave overtopping during the 100-year design life of the road, but the NTDD prevailed on us to reduce this to +5.5mPD which matched the carriageway levels in the adjacent reclamations for the new towns.

It was a particularly hot, dry summer in 1977 and there was a severe water shortage in Hong Kong. With the water supply restricted to 4 hours a day, all available pots and pans in our flat were filled with water when it was available. The flats in Silver Strand where we stayed lacked a swimming pool, but my wife had become friendly with some of the families in adjoining flats who were members of the United Services Recreation Club in Jordan Road in Kowloon who invited her and the boys to the pool at the club. At weekends if I were not working, we could swim in the sea off Silver Strand beach, or better, at the long Clearwater Bay beach where there were kiosks selling ice cream, soft drinks and cans of beer. As a change we sometimes went to Silvermine Bay on Lantau taking a ferry from Central. We also took the opportunity to drive around Kowloon, first to the tourist observation platform at the border looking into communist China across the twin security fences erected by the British Army. From there we could see people working in the paddies (the rural scene we were looking at has now been absorbed by Shenzhen, a city rivalling Hong Kong in size). Then we went to Tuen Mun, before turning back along the old Castle Peak Road to central Kowloon. In 1977 the roads throughout Hong Kong were all single carriageway as this was before the major colony-wide road building programme got properly underway. We often drove down the steep hill of Hiram's Highway and on past Pak Sha Wan and the yacht club in Hebe Haven to Sai Kung, near where we later lived for over six years. Sai Kung was and still is the major town in the north-east NT. In those days it was a small place with just two main streets and an attractive waterfront along which large seagoing junks from communist China sometimes moored. It was a typical Chinese town then as the influx of expats to village houses and new developments had barely started. When David and John were walking in the streets of Sai Kung with us, old Chinese ladies would cross the road specially to touch them on the head much to the boys' annoyance. The old ladies meant no harm and were merely seeking good luck from contact with their golden hair.

Sai Kung had a proud history as it had been the centre of the resistance to the brutal Japanese occupation of Hong Kong between 1941 and 1945. The resistance was organised through the Kowloon Brigade whose campaign of guerrilla warfare was so successful that Japanese detachments could only move around in large groups in the Sai Kung area. The Brigade was armed by free China with involvement of British agents forcing the Japanese to blockade Sai Kung against mainland junks and smaller craft landing weapons for use by the resistance. Sai Kung inhabitants remained sympathetic to communism up to the 1970s and beyond, a feeling not generally shared in Hong Kong when we were there.

That hot summer of 1977 also brought to a head discontent which had simmered between the Royal Hong Kong Police and the Independent Commission Against Corruption since the latter was established by the Governor, Sir Murray MacLehose in 1974 to deal with corruption. This had become widespread throughout government departments in the colony. It was almost institutional in the police with many British officers deeply involved along with Chinese officers in the district police stations who organised protection rackets and other illicit activities. Following ICAC investigations, several high-ranking officers were tried and convicted, the most famous of whom was the police chief, Peter Godber who had salted away over $4 million in overseas bank accounts and who fled to the UK, only to be brought back to the colony and sentenced to 4 years in prison. While most of the worst offenders, both British and Chinese, had been dealt with by 1977 prosecution of rank-and-file officers was still ongoing, many of whom resented what they saw over-zealous investigation methods employed by the ICAC. Shortly after my wife and family arrived the resentment boiled over with what was effectively a mutiny by policeman who stormed the police headquarters in Central. We watched all this on television in our flat in Silver Strand wondering where it was going to end. At one point, the colonial administration came close to calling on help from the British garrison, but the situation was defused without the need for this when the government announced a partial amnesty for minor offences committed before 1977. Throughout the period of police unrest, which in other places might have led to wider insurrection, life in Kowloon-side had gone on as if nothing were happening. Even the police continued their normal business as I found out one day when I was issued with a $30 dollar parking penalty. As he handed me the ticket, the young policeman politely explained in excellent English (he had the red shoulder-flash which signified his proficiency in the language) that I really should not park on the pavement at the electrical sub-station at the bend in Austin Avenue, and helpfully suggested alternative legal parking places nearby.

In mid-September when the feasibility study was completed, a report on the work with recommendations for construction was submitted to NTDD and we awaited approval to commence detailed design. Eventually, this was given, but the design moved forward in fits and starts as the Hong Kong government first urged us to get the project out to tender as soon as possible, then paused it for financial reasons and finally accelerated it again with the addition of an interchange and other road connections, adding four more bridges and requiring the relocation of two ferry piers. In the autumn of 1977, however, the firm won three other jobs, two in Thailand and one in Sabah in east Malaysia and I worked on all of them in their design phases, mostly in the countries in which the projects were located as well as spending a lot of time in Hong Kong.

We were joined in Hong Kong by Tony Bowley, another road engineer who was to become a great friend over the years up to his untimely death seventeen years later. When the recommendations of the feasibility study were finally accepted by the NTDD with the various additions, detailed design was commenced. A key feature of the scheme was the design of the reclamation and marine embankment in Tolo Harbour, in depths of up to 5m of water over a seabed of marine clay. As in much of Hong Kong's inshore waters, the seabed soils consist of soft marine clays and silts which had been deposited over time and which overlay residual soil and alluvium from earlier weathering processes. Embankment construction on the marine clay needs careful consideration to avoid excessive settlement. Complete removal of the clay by dredging would avoid any settlement issues, but an appreciable saving in cost could be obtained by only partial removal provided the settlement from the weight of the embankment on the remaining clay would be tolerable. There was an average depth of clay of 8m along the route and following extensive site investigation and geotechnical calculations it was decided to remove 5m of the clay, except for one area which was particularly susceptible to settlement where 6m would be removed. After dredging had been completed a submersible fill which could consolidate under its own weight would be placed up to around high water and then capped by a compactable fill to the final formation level for the road carriageways. Placing of the filling in two areas of reclamation close to the road would have to proceed carefully to avoid trapping "mud-waves" which might form in the remaining depths of soft clays. This had happened in other reclamations in Hong Kong and where unpredictable settlements had arisen. In areas where piles would be driven for the bridges and other structures the maximum size of rock in the filling would be restricted to 100mm so as not to impede driving of the piles.

The bridges were designed to be supported on steel bearing piles driven to weathered volcanic rocks below the clay and alluvium. This would ensure there would be virtually no settlement of the bridges themselves. In addition to carrying the weight of the bridge structures and live loading from traffic on the bridge decks, the piles had to be designed for negative skin friction on the pile shafts as the clay layers consolidated and filling over them moved down around the piles. Run-on slabs were provided at all bridge abutments to smooth the transition between bridge and the embankment on the approach which would suffer some settlement. In the event settlements in the embankments were considerably less than the geotechnical engineers predicted.

When post-tensioned deck beams are used, rectangular end blocks are required to anchor the prestressing cables. As the deck beams were to be placed side by side, spaces of only 200mm would be left between these end blocks after erection of the beams. This is insufficient to lap reinforcing bars needed to resist torsion effects in the end diaphragms and a transverse prestressing system was designed for the ends of each deck to clamp the end blocks together. In the event this worked well.

Two of the bridges which had been added were part of an interchange near Tai Po where the Trunk Road swung west of the town towards Fanling. As these bridges were on tight curves they were not suited to precast beams. One of them was designed with a continuous flat slab reinforced concrete deck on groups of circular piers and the other as a prestressed concrete box girder. There was also a footbridge at the Sha Tin end of the road which carried the cycle track over the carriageways. I designed this as a slender single span structure using in situ post-tensioned concrete, constructed on falsework prior to stressing the cables. To fit in the west approach span to this bridge in the space available we had to squeeze it very close to the Kowloon Canton Railway (KCR) embankment. This gave some interesting problems in the construction stage.

While I was finalising the design of the bridges there was a change in management of the Hong Kong office and an ex-RPT engineer called Ted Foster took over responsibility for the office. He had moved to Australia with RPT when they opened an Australian office and had decided to make his home there. However, Peter Fraenkel persuaded him to take over running our Hong Kong office, recognising belatedly that opportunities were being missed and Ted was much more switched on as far as promotion was concerned.

As the programme for finalising the detailed design of the Trunk Road bridges was subject to delays while the necessary financing was agreed within the Hong Kong government, I was able to work on the three other jobs which the firm had won in the Far East, as well as retaining an involvement in Hong Kong. The detailed design of the two projects in Thailand overlapped for me, followed by the project in Malaysia, after which I returned to Hong Kong full time.

Dock Gates for the Thai Navy

The dock gates were part of a large project to design a new dockyard for the Thai navy. There was a small naval dockyard on the Chao Phraya River in Bangkok itself, but it had little room for expansion for the new vessels which the Thai government considered were required for potential regional conflicts. In the 1970s the Vietnam War was still in progress, but it was clear the Americans were looking for a way out. Vietnam, Cambodia and Laos were all likely to be under the influence of communist China and as such possible threats to Thailand. The Thais had bought two frigates from the Royal Navy along with patrol vessels and a new dockyard was seen to be essential to maintain them.

John Stanbury's contacts in the Ministry of Defence from his earlier career may have helped with introductions to the admirals in the Thai navy responsible for procurement, and one way or another, the firm won the consultancy contact to design the new dockyard. The site was at Pom Prachul at the mouth of the river 20km downstream from Bangkok. It was about as unsuitable for construction of a dockyard as could be, as it was on soft ground in a mangrove swamp, but nowhere else would have been any better as Bangkok sits on an extensive river flood plain with similar poor ground conditions all round. At least the chosen site had access to deep water in the Gulf of Thailand without the need to dredge an approach channel and it was quite close to the Thai navy headquarters in the city. The dockyard was to have 700m of riverside jetties, a semi-tidal holding basin serving two dry docks and fitting-out quays, a 650t ship-lift, workshops and stores and living quarters for staff. A design team was established in which I had the job of designing the two dry dock gates. There was also an entrance gate to the basin and an emergency caisson for use if any of the gates failed or had to be removed for repairs. Unlike the dry dock gates which were intended to be constructed to the Engineer's design, the entrance gate and the caisson were to be designed by the successful contractor against performance specifications.

Soft Bangkok Clay extends from the old capital Ayutthaya in the north, through Bangkok and down to the mouth of the river where the site was located. A marine clay with a depth of over 20m, it overlies stiffer fluvial clays and sands. It is so soft it cannot support structures or filling without uncontrolled settlement and all the facilities in the dockyard would need to be supported on piles. So soft in fact, I have seen precast concrete piles for new buildings sink 5m or more into the clay under their own weight when pitched end-on ready for driving. The twin drydocks located on one side of the basin would require to be constructed in deep cofferdams, involving significant engineering in the soft ground. The frigates had a beam of 13m and a draught of 6m, but the dry docks were designed to accommodate bigger vessels which might be acquired in the future and were to have an entrance width of 18m and a water depth over the sill of 9m which determined the size for the gates.

I opted to design the dry dock gates as buoyant flap gates. Operation of the gates would be semi-automatic and controlled by a combination of compressed air and valves without winches or other mechanical devices, except as emergency backup. I cannot claim this design concept as my idea, however, as I copied it from an earlier design by Maunsell for a gate at No. 4 Dry Dock in Malta (in the days when the dockyard there was still a Royal Navy base and a well-written and informative paper on the Malta gate had been published in the ICE Proceedings). Although the concept was not new, I think I improved it with a few tweaks of my own. Each gate would be designed with an upper steel box girder to act as a buoyancy chamber and span across the entrance with a stiffened steel skin plate extending down below the box to hinges at the sill. When the dock was pumped out with the gate in position, water pressure would force it against the sides of the dock with rubber seals on the bearing surfaces. The Malta gate had two fixed hinges at the sill, and I used two hinge points as well. However, I was concerned that unless these hinges were installed to extremely tight tolerances, they would tend to prevent the gates fitting snugly against the sill and the sides near the bottom of the gates so that leakage could occur. Maunsell had got over this to some extent by intentionally using a loose fit of the hinge pins in the bearings. I thought that might lead to excessive wear in the longer term and instead designed a double hinge or linkage arrangement to ensure that the gates would not be restrained in any way from complete contact with the bearing faces, both along the sill and on the sides. The arrangement was extremely successful and there was scarcely any leakage to be seen inside the dock when it was pumped out with the gates closed.

To lower a gate, valves would be opened in the box girder to partially flood it. With the loss of buoyancy, the gate would rotate about its hinges and settle into a recess in the dock apron. Raising the gate again would only require blowing out the buoyancy chamber with compressed air and the gate would come up into its operating position.

The structural design for this concept was extremely simple. My initial hand calculations were augmented by a grillage analysis, but the bending moments and shear from the grillage model scarcely differed from the hand analysis. I used hard rubber seals at the sills with softer seals on the sides of the dock except at the box girder contact points where hard seals were used. This ensured that the skin plate in each gate spanned from the box girder at the top to the sill at the bottom with little load picked up on the soft sides and the moments and shears in it were therefore virtually statically determinant. It was the same for the box girder which was the main load bearing part of the system and spanned as a simple beam in bending between the hard seals on dock walls. The important thing was to get the geometry right so that the gates pressed against the sides naturally when they were raised with the dock flooded, without the need to be drawn against the walls or latched to keep them in position. This was achieved by ensuring the separate lines of action of the upward thrust due to buoyancy and the submerged weight of the gates acting downwards through the hinges created a moment (turning effect) at the hinges. With the geometry correctly set up and thanks to the linkage arrangement at the hinges the gates fitted against the sides snugly.

As always in new designs, there was one issue which could have been troublesome. The water at the mouth of the river had a high silt content and in the confined basin of the dockyard with little in the way of currents the silt could settle out. This was allowed for in the scheme by accepting the need for regular maintenance dredging in the basin to preserve under-keel clearance for vessels. At the dry dock entrances the silt could also settle out and here it would not be easily dredged and could build up on the gates when they were open and lying in their recesses on the apron. If silt built up excessively its weight could prevent a gate rising when the buoyancy tanks were blown. We therefore arranged for additional compressed air lines to be provided, designed to blow silt clear prior to gate operation and a second air system was provided to clear silt from the recesses. These systems turned out to be successful and the gates operated without problems for many years after installation. We had a scare at the opening ceremony of the dockyard, however, as described later.

The construction contract was won by a joint venture of Zublin, a long-established German contractor from Stuttgart, and Christiani & Nielsen, a British/Danish company. Having designed the gates, I dropped out of the project for some time, returning to Hong Kong on the Trunk Road bridges and spending some time in Malaysia on one of the other South-east Asian projects, the Sabah Rural Trunk Roads, which was starting up. I was not involved with the dockyard again until early 1981. By that time construction was well underway and the contractors had arranged to have the steel gates fabricated in a shipyard in Singapore.

Ideally, we should have had a full-time steelwork inspector in the fabrication shop, but to save money I was asked to fill in for six weeks when the work was starting up to make sure things were set up properly. I was not a steelwork inspector, but I knew enough about steelwork fabrication from Ballachulish Bridge and Tilbury Docks Floodgate to get by, checking mill certificates for the plating, reviewing welding procedures and flame cutting trials and going through the fabrication programme with the production manager with whom I got quite friendly. The shipyard management and workers were competent, and my presence made little difference to the quality of the work which was excellent. We had produced a set of fully detailed drawings for the dry dock gates in London and the fabricators paid me the compliment of working directly from these drawings without preparing separate shop drawings as is normal practice. Where there was any ambiguity or lack of information on the drawings, I would mark them up with the necessary information and all went very smoothly. I stayed in the Cockpit Hotel, much beloved as by old Malaya hands Geoff Nicklin and John Tainsh who had worked in Kuala Lumpur in the 1950s and always stayed at the Cockpit when they visited Singapore. It was one of the original hotels in Singapore dating back to the 1840s and had a faded colonial ambience which I liked. Every day I was picked up at 8.30am and driven to the shipyard which was in Jurong and then brought back to the Cockpit again by car in the evening. I had plenty time to explore Singapore, but as I was on my own, I did become rather bored. However, the shipyard management were hospitable and organised dinners in various Chinese restaurants around the town and a memorable trip across the Causeway to Johore Bahru to see two English strippers perform in a place packed with excited Muslim Malays.

The Thai Navy Dockyard was formally opened in 1982 by King Bhumibol (later conferred Bhumibol the Great) from whom I received an invitation to attend the ceremony in the company of Peter Fraenkel, John Stanbury, Eric Phillips (the ER) and some of the site staff. I travelled over from Hong Kong for the opening ceremony, mitigating the cost to the firm to some extent by doing a little work on another Thai project while I was there. On the day of the ceremony, we were lined up along with staff from the contractors on one side of the dry docks. These were flooded with the frigates berthed in them and with both gates lying in their recesses on the floor of the dock. It was a hot, hazy May afternoon just before the rains and the King was scheduled to arrive with his entourage at 3pm. The most symbolic part of the opening ceremony was to be the raising the dry dock gates which would signify the facility was ready for operation in its function of providing a secure base for maintenance of the frigates.

On the previous day Eric and I had arranged for the gates to be raised and lowered several times and everything had worked perfectly. It took exactly one minute and twenty seconds from activating the compressed air system and pressing the raising button for the top of the first gate to break the water surface and about 10 seconds more with the second gate. All that was required on the day was for the operator to arm the systems and press the buttons and we were confident nothing could go wrong. The King duly arrived and was met by several admirals who presented him to a guard of honour and led him across to the dry docks. Buddhist monks then conducted a short ritual, presumably blessing the dockyard and praying for its successful use, and there was an expectant silence with all eyes on the water at the dock entrances. A minute or so earlier we had seen the operator in his cabin move to press the buttons and we expected to see the leading edge of the first gate break the surface any second. One minute twenty seconds passed, but nothing happened, then two minutes and still nothing. Now there was some shuffling and low comments amongst the onlookers. What could have gone wrong? An unexpected build-up of silt or an electrical failure? I looked at Eric and we both silently mouthed the same oath. Then, suddenly there was a swirl in the water in No.1 Dry Dock and the upper corner of the box girder of the gate appeared, followed by the gate in No.2 Dock and both gates moved smoothly and steadily through the final parts of their arcs to close sweetly in the entrances. There was applause from the guests and the King raised his hand in salute. I felt in need of a stiff drink.

It transpired that the operator had not actually pressed the buttons. He had merely put his fingers on them to be ready when the monks' chants were coming to an end and had misjudged the timing. In any event the operation of the gates was judged a great success with congratulations all round. We had some small eats and champagne on the site and were each given a bronze medal marking the occasion and the contractors arranged a memorable party that evening in the old Erawan Hotel. I flew back to Hong Kong the next day, suffering from a hangover, but nonetheless flushed with the success of the gates.

Rama IX Bridge, Bangkok

The project began with a feasibility study for a crossing of the Chao Phraya River connecting Bangkok on the east side of the river to Thonburi on the west. This was the third section of the first phase of an urban expressway system in Bangkok, running roughly east to west through the city. The crossing was to be either in a tunnel or a high-level bridge with a clear span over the river for navigation. PFP had won the consultancy commission in January 1980 as lead consultant in a consortium which included Dr Ing. Helmut Homberg of Germany (specifically for a bridge), Parson Brinckerhoff International Inc. of New York (PBI were tunnelling specialists) and a local Thai consultant National engineering Consultants Ltd (NECCO). I had been involved in discussions with Dr Homberg in London when we were preparing the proposal and, in fact, it was me who introduced Homberg to PFP with the help of David Wright who was working with him on Crouch & Hogg's Kessock Bridge in Scotland at the time. However, I was in Hong Kong when the feasibility study was carried out and had little involvement in it. The study was completed in October 1980 and recommended a high-level cable-stayed bridge as a tunnel would have been very costly in the soft clay ground. Dr Homberg gave the impression of being the almost archetypical arrogant German we see in caricatures, which to some extent he was, but he was also a brilliant engineer. As he was not slow to point out, he had been the designer of roughly half the cable-stayed bridges constructed in Europe after the war. The bridge he designed for Bangkok drew on his great experience and had an elegant simplicity with a single, central plane of stay cables fanning out from towers on either side of the river.

The design stage of the bridge crossing started in January 1981 with a full site investigation for the at-grade roads and the foundations for the elevated structures to augment a more general SI which had been carried out for the feasibility study. I was to be responsible for the final detailed design of the approach viaducts and interchange bridges on either side of the river. This work was carried out in a design office set up in Bangkok once SI results were available. The main cable-stayed bridge was to be designed by Dr Homberg in his office in Germany with help from our London office on the foundations. Hans Eilenberg was project manager for the design, pulling together Homberg's work with mine into a tender package. I arrived in Bangkok in March with my wife and family scheduled to join me in June. Reviewing what I had been done in the study I immediately had misgivings on the form of approach bridges which had been recommended. Unfortunately, it was too late to change the concept.

The study had recommended a series of simply supported spans of 50m on each side of the river, rising from ground to the main bridge deck at an elevation of 45m. While the main bridge carried the east-bound and west-bound traffic lanes of the highway, six lanes in all, on a single full-width deck section, separated only by a central barrier, the approaches comprised two separate structures each with three traffic lanes with a one metre air gap between them. I had no argument with this arrangement, but I was not happy with the simply supported concept for the approaches for which we had geotechnical engineers to thank and would have much preferred to make the approach decks continuous over the piers. The geotechnical engineers' view, without any real evidence to back it up, had been that there would be differential settlement of the piers as a result of deep-well pumping for water supply below the Bangkok Clay. The aquifer would be replenished close to the river, they said, and the settlement would become progressively greater the further from the river one went. Thus, simply supported spans had to be used, as continuous decks would suffer cracking. However, nobody had taken the trouble to draw out a ground profile for the worst-case settlement scenario which would have shown that the small differential settlement between piers 50m apart would not have troubled a slender continuous bridge deck, especially when it was prestressed. Such a deck would have been easier to construct and would have had a greater intrinsic collapse strength than a series of simply supported spans. Overloaded continuous bridges can only collapse after formation of several plastic hinges, whereas simply supported spans need only one, at the centre of the span. On aesthetic grounds, which I considered important, a relatively slender continuous bridge would have had a better appearance than the deeper decks needed with simply supported construction. In fact, continuous bridges for other crossings of the river were built subsequently in Bangkok without any problems, proving my point. But the die was cast before I got involved and I had to make the best of it.

Dr Homberg played no direct part in the design of the approaches, just as I was not involved in the design of the main bridge, but he helped me in one respect. He provided me with an English translation of a monograph he had published in 1973 entitled *Platten Mit Zwei Stegen* (in English: Double Webbed Slabs). This comprised figures and tables of influence surfaces for bending moments from live loading for various configurations of double tee or double web bridge decks, derived from classical slab bending theory. With these to hand I adopted a double tee bridge deck configuration and the monograph proved extremely useful for the design. I still have the copy he gave me.

My design based on his charts was for 15m wide decks with a basic slab thickness of 0.3m and two 3.5m deep by 0.9m wide ribs spaced 6m apart. The decks were prestressed with 10 cables in each rib. Each cable comprised a total of 19 low-relaxation super grade strands in a draped profile between the end anchorages, designed to be stressed from one end. The specified ultimate strength of each cable was 4950kN. I made some slight modifications to Dr Homberg's chart values to account for shear lag effects in the wide deck slab between the ribs, adapting the latest BS 5400 rules on shear lag for steel composite bridges. I felt it was a clean, simple design and for a series of simply supported spans as efficient in the use of materials as possible.

I also designed the piers and their piled foundations. As the approach bridge spans were all simply supported, bearings were needed under the seating of the ribs at the ends of each span, two on one side of the pier and two on the other. To accommodate these sets of bearings the piers needed to be at least 4m across in the line of the bridge. This was another unfortunate consequence of the simply supported configuration as the piers could not look other than rather heavy and ungainly. If the decks had been continuous only two bearings would have been needed on each pier cap and more slender piers could have been used. As they were going to be of large section to accommodate the bearings, I opted to use single central piers under each carriageway deck and made them each 7m wide, just enough for seating of bearings with the ribs spaced 6m apart. Vertical grooves were provided on the four faces of each pier to break up what would otherwise have been flat concrete panels. To cut down on concrete and self-weight, I designed the piers as hollow columns with internal ladders leading up to pier caps where there were manholes for access to the bearings.

The piles in the foundations were designed to be driven to a hard clayey sand layer below the soft Bangkok Clay at a depth of around 25m with vertical prestressed concrete piles for the vertical loading and steel piles raked at 1 in 4 to resist horizontal forces. It was supposed to be a cardinal sin to mix steel and concrete piles in the same foundation, but I ignored this despite opposition from geotechnical engineers. Their objection was that settlements in the soft clay would bear down on the steel raking piles and cause them to fail. This was nonsense. There was no significant surcharge to consolidate the natural ground and the steel piles would be relatively flexible if any local consolidation did occur. Fortunately, Hans backed me up in my decision.

The local consultant NECCO provided office accommodation, some assistant engineers and a large pool of draftsmen. They also arranged "entertainments" for us such as visits to their favourite massage parlours and golf outings. I did not take up their invitations to any of the former, I but did go on the golf outings which were quite unlike any office golf outings I had been on in the UK where normal golf etiquette was generally observed. In Thailand, however, a much more relaxed attitude prevailed. On the second outing, I was playing in a four-ball with hired clubs, and we were making painfully slow and erratic progress, with the NECCO golfers hacking their way from one side to the other up the fairways. Eventually we caught up with a threesome of Japanese golfers even slower than ourselves. Rather than inviting us to pass through, the Japanese politely asked if they could join us and did so slowing us down even more. I had started out playing reasonably well and had a knowledgeable young caddy who spoke good English and unerringly selected the correct clubs for me on the approach to each hole. However, my game started to deteriorate in the chaos around me with seven golfers and their caddies, strung out up and down the fairways, like so many ants scurrying hither and thither, and I looked for some way to escape. As my caddy and I reached the twelfth green a long way ahead of the rest of our group I could see a large gap had opened in front of us with the nearest other golfers at least three holes ahead. The answer therefore was to move on without waiting on the rest of the party to catch up and just play the remaining holes on my own, or even better, with the caddy, as I guessed he would be a useful golfer as caddies usually are. First, I signalled back to the NECCO partner who had organised the outing that I was packing up, miming a headache from the sun. Then off we went quickly to the thirteenth tee which was out of sight of the twelfth green where I drove off and offered the driver to the caddy to do likewise. He did so with a smile, hitting a booming drive down the centre of the fairway at least twenty yards beyond mine. My golf improved again, and while I was no match for the caddy who beat me 3 and 2 over the six holes we played, it was by far the most enjoyable part of the round. Back at the clubhouse I returned the hired clubs, tipped my caddy 100 baht and headed for the veranda where I had plenty time for a few pints of Singha beer before our party came in. They voiced some concern at my headache and the possibility of sun stroke, but I assured them that my problem had simply been dehydration which I had quickly remedied with an appropriate intake of fluids.

Once I had completed the basic design of all the main parts of the approach bridges, I went back to the UK for three weeks, to help pack up and arrange for shipping of some of our effects to the Far East. After Bangkok, we were going to Malaysia where I was to be responsible for the detailed design of bridges on the Sabah roads project and then eventually back to Hong Kong where I was to become an associate partner in the business there. As we expected to be overseas as a family for some years, agents had to be engaged to manage letting of our house in Scotland. While we were making these arrangements a sad event occurred. My friend from Fairhurst days, David Wright, had been diagnosed with mesothelioma about six months earlier, and he died shortly before we left for the Far East.

Now in Bangkok as a family, we stayed in the firm's flat in Soi Lang Suan. I was very familiar with it having stayed there in 1977 with Peter French and when I was on my own at the start of the design process. We had a full-time maid, *Nit*, the name shortened from *Nit Noi*, meaning little one in Thai, although she was quite tall. She was very intelligent, speaking several languages fluently, including Japanese (she had worked for a Japanese family for four years) and an excellent cook. But she was also a law unto herself to some extent and needed diplomatic handling at times. She was a de facto *mama san* for the maids in the block of flats we were in and organised them in various ways. When a vacancy for a maid arose, she would bring in a suitable girl from up country and if she had some of her own business to attend to, she would press one of the other maids into service on her behalf. Once we wanted to have some friends around for a small dinner party, but Margo was concerned we did not have enough plates and cutlery. However, *Nit* assured her there would no problem in that regard and simply took what was needed from other flats in the block. Our meal was served by several maids from those flats with *Nit* in command in the kitchen, only appearing in the dining room occasionally to receive well-earned plaudits for the excellence of her cooking. The other maids did the washing up and the plates and cutlery were returned to their rightful owners who were none the wiser on what had happened.

I had joined the British Club on earlier visits to Bangkok. The club was a haven of peace and sanity out of the frantic Bangkok traffic in a lane off Silom Road, not far from the Oriental Hotel. It had a lovely swimming pool, a good restaurant and was altogether a very pleasant, well-run place. Not surprisingly, Margo and the boys spent a lot of time in the club. There seemed to be a lot of children in the club at that time, many of them boys, mainly British, but also of other nationalities and David and John soon made friends. The lifeguards had their work cut out trying to exercise some control of swarms of boys running wild around the pool, as parents like us paid little attention to them.

I had the use of a car with a driver for day-to-day travel around Bangkok, and on some weekends, I would take the car and we would drive down to the south of Thailand, to Pattaya on the east side of the Gulf or to Hua Hin on the west side. At that time there was only one hotel in Hua Hin, unlike now when it is a major package holiday destination. This was the Railway Hotel (Hua Hin lies on the line to the south and Malaysia) and it was rather run down. We stayed in a chalet in the gardens, probably a mistake as there were clouds of mosquitos all around and Margo was quite badly bitten. There were two good restaurants in town, however, which compensated to some extent.

There was a strange occurrence in early August that year. In 1981 Cambodia was turmoil and there were frequent armed incursions into Thai territory by the successors to the Khmer Rouge. Thailand maintained a large military presence in the border areas and consequently tensions there were high. A group of high-ranking officers in the Thai army had visited the border and were flying back to Bangkok in a helicopter when it crashed in a thunderstorm killing everyone on board. The next day at the time of the crash there was a severe thunderstorm in Bangkok. Thunderstorms are common in Thailand in August as it is the peak month of the rainy season, but the severity of this storm was exceptional. The same thing happened at the time of the crash the next day and the next, each storm more violent than the one before. As *Nit* and Margo discussed the storms in our flat in Soi Lang Suan, there was no doubt in *Nit's* mind that the spirits of the army officers were involved, and storms would occur each day until funerals were held with proper Buddhist rituals so that their spirits could be released from the earth. On the day of the funerals the storm began around lunchtime and thunder and lightning became almost continuous. It was undoubtedly the most violent thunderstorm I ever witnessed. I was on the fifth floor of NECCO's office building at a window and saw bolts of lightning streaking down to hit the ground all around, each one accompanied by a loud hiss, followed immediately by a deafening crash of thunder. But quite suddenly, the storm abated, the rain eased off and the hot sun came out. Back at the flat *Nit* said the funeral ceremonies had all been completed and the spirits had departed in peace. There were no further storms and the next day things returned to normal.

Despite my reservations on the concepts for the approaches, the detailed design progressed smoothly. I had an engineer from London called Chris Kendall to help me and we soon had all the details worked out. By then we had moved on from slide rules and used HP 41C programmable calculators which were quite useful as many of the routines from BS5400 could be set up on them saving on calculation time. Our main problem was not the design itself, but getting the drawings completed. These included a method of erection which would suit the design requirements for the permanent works with no overstressing of the structures. I was never sure if it were a good thing to include such drawings in a tender set as the Engineer might be blamed if things went wrong, but the method we indicated had been used on double tee decks in Europe and we were confident it would work. It involved construction by the contractor of a special travelling steel structure, capable of being launched forward span by span. This would be used to support steel formwork for the concrete deck. After a span had been cast, it would be stressed, and the travelling structure would be jacked down on to rollers on the pier tops before moving forward to construct the next span. I produced an outline design and drawings for such a system, although the contractor would not be precluded from using another method if he so desired, subject to approval by the Engineer. In the event, the system I indicated was used by the successful contractor with few changes. With a monumental effort by NECCO's draftsman all the tender drawings were completed on programme, including those of the erection scheme. With the design complete, we packed up and headed off to Malaysia.

Sabah Rural Trunk Roads

The project in Sabah involved four separate roads, two in the west, one from east to west across Sabah and one in the east. An economic case had to be made for each of them and, in the event, neither of the roads in the west showed sufficient positive benefit in terms of the World Bank cost/benefit criteria for lending and were dropped from the scheme. The other two roads did proceed, however, and were eventually constructed. These were from Sandakan west to Telupid and from Sandakan south to Lahad Datu, with six bridges on the former route and five on the latter. Two of the bridges on the Sandakan to Lahad Datu Road were across major rivers, requiring multi-span bridges and one bridge on the Sandakan to Telupid Road was also multi-span. All the others were single span structures.

Before independence from Britain, Sabah was called British North Borneo and is often referred to as "the land below the winds", meaning typhoons which invariably track to the north. It had become independent in 1963, and joined the Malaysian Federation, along with Sarawak, while the Sultanate of Brunei became independent outside of the new Malaysia. Malays make up only a small percentage of the population of Sabah, being outnumbered by Kadazans, Chinese, and many other ethnic groupings who are largely Christian, in contrast to the Sunni Muslim Malays. Sabah is mountainous with the Crocker range running north to south and the highest mountains in the north-west of the state reaching over 4,000 metres in Mount Kinabalu. It was still heavily forested when we won the roads job, although logging activities were already stripping out large areas of the primary rainforest to be replaced with oil palm plantations. Despite this, the forest had the vast, brooding presence most people associate with Borneo. One always expected to see the famed "wild man" when *Ulu* (up-river in Malay vernacular). Whether the wild man was human or animal in the shape of the orangutan I was never sure.

I was responsible for design of the bridges, starting in the feasibility study for the project in 1980 when I developed the concept designs, and a year later 1981, after completing the design of the Rama IX Bridge approach structures in Bangkok, when I carried out the detailed design. For this project we had teamed up with a consulting engineering firm from Peninsula Malaysia called Minco, a competent organisation and the largest civil engineering consultant in Malaysia in 1980 with 500 employees. The firm had been founded in 1962 by P. Ganendra who had been a civil engineer in the British-run railways and was known to Geoff Nicklin and John Tainsh who had both worked in KL in the 1950s. We knew him as "Gerry, but I'll refer to him as Ganendra for the purpose of the memoir. He was Singhalese, and very astute in business matters as the success of his company demonstrated. You would have had to be up very early in the morning to get the better of him. To go with his astuteness, he had a droll turn of phrase, and I came to like him a lot. There was also a local consultant firm from Sabah in the group, KKK, who arranged an office for the feasibility study and provided some assistance on the ground. This office was in the capital of Sabah, Kota Kinabalu, invariably referred to as KK, and formerly called Jesselton prior to independence from Britain. As KK had been extensively damaged by bombing during the war and largely rebuilt in the 1950s, it lacked the character of some of the towns in Peninsula Malaysia. Nevertheless, it was an extremely pleasant place to stay with the KK Club above the town for squash and drinks, a yacht club on Tanjong Aru Beach near the town and coral islands just offshore for snorkelling. There is a stunning backdrop of mountains to the north, rising to the peak of Mount Kinabalu which is usually visible in the mornings before clouds build up on the mountain in the afternoon. It was also a very peaceful place as crime in Sabah was virtually non-existent at that time. As the study got underway, I spent some months in KK on an unaccompanied basis before my family joined me in June of 1980.

During the period when I was on my own, I travelled widely in Sabah visiting all rivers to be crossed and surveying the most suitable locations for the bridges. At the feasibility study stage, the western routes were still under consideration. For the bridges on these, I went as far as the border with Sarawak in the south and up to Kota Belud in the north of Sabah. The most interest lay in the routes from Sandakan in the east of Sabah, however, and to get to that part of the state it was necessary to take an internal flight to either Sandakan or Lahad Datu, passing close to the peak of Mount Kinabalu on the way. For investigating bridge requirements on these routes, I stayed in a pleasant hotel just outside Sandakan and had a Toyota Land Cruiser with a local driver. Sandakan itself was very Chinese in character with covered "five-foot ways" in front of the shop houses, as in old parts of Singapore, and boasted some excellent seafood restaurants.

From the Sandakan base we could cover the entire route to Telupid in a day. This was the first part of the infamous forced march from Sandakan to Ranau during the Japanese occupation of Sabah when over 2000 Australian and British POWs died. Some of the route ran along the valley of a river which was full of crocodiles where the best site for one of the river crossings was just downstream from a small kampong. My driver enlisted help from two young men in the kampong who took me down the river in a dug-out canoe to see the potential crossing point. As they paddled back up again, I saw several crocodiles slither off the muddy banks. The young men saw them too and called out "*Buaya*" and took great delight inducing a violent rolling motion in the canoe—easily done in a dug-out with no keel. Fortunately, they knew how far go without overturning us and we made it back to the kampong. To compensate for my scare some delightful young ladies in hijabs appeared with refreshments including green coconuts which one of the young men deftly cut open with a *parang* to drain out the milk for me. All good fun I thought, at least when one is safely back on the bank.

The location of one of the other bridges on this route was more difficult to reach as it was on a smaller non-navigable river in a thickly forested area. To get to it one had to walk about one mile through the forest and again my driver enlisted help from a nearby kampong to guide me to the river. Part of our route to the bridge site was through the primary rainforest with a dense tree canopy over one hundred feet overhead. This part was easy walking as the large trees were well spaced although gibbons screeched from the treetops and broke off branches to throw at us as we passed below them. The diffuse light reaching the forest floor through the tree canopy overhead, typical of the primary rainforest, was insufficient to support much undergrowth and we made good progress. Further on, however, the forest had been cleared in the shifting agriculture practice still employed around some of the kampongs, then allowed to revert to jungle again when the cultivated area was abandoned, and a new area was cleared. The removal of the tree canopy and the unrestricted sunlight had resulted in a rapid regrowth of vegetation which was almost impenetrable in places. My guides had to cut a path for us with their *parangs* and progress was slow until we came near the river where the secondary growth opened out again. Fortunately, we reached the river at a straight reach suitable for a bridge crossing as it would have been difficult to move up or down the bank to find another site and I was able to estimate the width of the river to obtain the necessary bridge span. On our walk back to the kampong I took a series of compass bearings, so that the crossing point could be located by the road engineers using reciprocal bearings from the kampong when they were designing the road alignment.

To reach the southern part of the other route from Sandakan beyond the Sungai Kinabatangan, an overnight stay in Lahad Datu was required, although there was an airstrip near the town which one could fly to if travelling direct from KK. At that time, Lahad Datu was a sleepy little place. Louis Akroyd, the geotechnical specialist I worked with in Nigeria, came out to Sabah for a few weeks and famously described it as "a one-horse town without the horse". But like small towns all over Malaysia, both east and west, action was to be had if so wished. There was a somewhat indifferent hotel we stayed in and if one went to the bar before having a meal, the *mama san* would appear and enquire if the gentleman would be "dancing" that evening. The dances she had in mind were the jig-a-jig or mattress polka and she was asking politely if one was seeking a girl for the night. She would not be in any way put out if the offer was declined equally politely and would signal to a waitress to take an order for beer.

There was a Standard Chartered Bank in Lahad Datu where money could be drawn if needed for a stay in the town. Some years later in 1985 when the road and bridges were under construction, the bank was the scene of a violent hold-up by pirates from an island in the southern Philippines not far offshore. This resulted in the deaths of twenty people who were in the bank and its vicinity at time, and although the pirates were engaged by police, most of them got away on their fast boats to Philippine territory. Retribution was visited on the island about a month later by Malaysian forces in a major operation involving the Malaysian navy, and most of the male inhabitants of the island were killed.

Unlike Northern Nigeria, the rivers to be crossed were usually running nearly full as Sabah has regular rainfall in NE and SW monsoons. Concrete aggregates are not abundant and early on I decided to design all the bridges as steel structures of one form or other. At times of heavy rains, river flow velocities could be rapid and where the soils of the beds were mobile, scour of foundations was a potential problem. The answer was to locate the bridges on straight reaches of the rivers, as far as possible, and to keep the abutments well clear of the banks with stone-filled gabion protection on both sides of the abutment structures. Any river piers were best supported on piles to guard against local scour around the pier. As with the design of river bridges in Nigeria, I carried out basic hydrological calculations for maximum flood discharge on each river to be crossed and ensured adequate waterway area for this at the bridge, adding a further metre of freeboard to the underside of the deck for good luck. By the time my wife and family arrived in Sabah I had a pretty good idea of the bridging requirements at each river and could proceed with concept design in the KK office without the need for further site visits.

As a family we stayed in idyllic surroundings in the old Borneo Hotel about two miles out of KK close to the beach at Tanjong Aru, in two well-appointed chalets beside the swimming pool. The hotel was owned and run by a Chinese lady who had a reputation for giving short shrift to complaints about the service or any other matter relating to the hotel, although we never had any cause for complaint during our stay. There was a Chinese restaurant in the hotel with tables on a pleasant veranda and a variety of places to eat in KK itself, including a somewhat rambunctious place called the "Diamond Restaurant" just out of town, much loved by the two boys. I had hired a Ford Consul for my wife and generally also had use of a short-wheelbase Toyota Land Cruiser for off-road trips to other beaches around KK, such as Dalit Beach, down a long track about 20 miles to the north and the epitome of a paradise beach in the tropics. When David and John arrived neither could swim properly, although David had had a course of lessons at home. This changed in a week when two other British children, a boy and a girl of similar age to our two, appeared with their parents, Pam and Richard Douthwaite, making their way back to Hong Kong from a holiday in Burma (this was before the name Myanmar came into use and quite an adventurous place for a holiday at that time). I did not know Richard then and did not actually meet him in the Borneo Hotel as I was working most of the time, but Margo became friendly with Pam, and I got to know Richard very well later when we played in the same squash team in Hong Kong. They were family members of the Kowloon Cricket Club with its good swimming pool, and the children were both excellent swimmers. Seeing them ploughing up and down the Borneo Hotel pool was too much for David and he taught himself to swim within a few days simply by emulating them, followed suit by

John a few days later. Within two weeks they were both snorkelling and diving down 20 feet inside the reef on one of the coral islands off KK.

I worked in the office on the conceptual design of the bridges and in preparing construction cost estimates for the cost/benefit analysis the World Bank required in the study. For this we had engaged Coopers & Lybrand as sub consultants, and they provided two transport economists to work with us. One of these was Mike Dyson with whom I shared an office. We did not get on to start with and rubbed each other up the wrong way for a time. Mike could be rather blunt in his dealings with people (perhaps a bit like David Wright) and I probably resented the fact he got paid a lot more than me for what I considered to be an inferior job. But we soon became quite friendly. He told me he had played a few games at full back for Hull Kingston Rovers before he had broken his ankle and had other health issues. He certainly knew all the songs, even filling me in with some words I had forgotten, but I could not to persuade him to play in the football team we had formed for games against local sides. I found out why this was later.

With the information I had got from the surveys and much better mapping than had been available in Nigeria, the concept design of the bridges was relatively straightforward, except for the bridges for the two large rivers on the Sandakan to Lahad Datu Road.

The first of these was the Sungai Kinabatangan about 100km south of Sandakan. At the crossing site the river was around 100m wide, and a chain driven ferry was in operation. The Kinabatangan is the largest river in Sabah. It rises in the mountains of the Crocker Range in south-west Sabah, flows east through the rain forest and discharges into the Sulu Sea near Sandakan. With its extensive catchment area, it is subject to large rises in water level in the rainy season in December and January during the NE monsoon, and even in the dry season, rainstorms in the catchment can cause its level to rise quickly and fall again rather more slowly when the rain stops. These fluctuations in water level caused problems in ferry operations as the banks of the river could become unstable giving vehicles difficulty getting on and off the ferry. The ferry also broke down regularly and could be out of commission for days at a time.

Obviously, bridge piers were best kept clear of the river channel and a main span of at least 140m was indicated. I favoured a 3-span bridge and with concrete aggregates in short supply a steel structure was preferred. This brought Ballachulish Bridge to mind and knew I could easily modify (and improve) the concept to suit the Kinabatangan crossing. In Ballachulish Bridge the trusses were in "N" form with vertical members at each node. I had always thought this a bit fussy, although CBE liked it as they claimed it simplified erection. The classic "W" Warren Girder form of truss to my eye looked much better with cleaner lines and I thought the erection issue could be overcome with a little ingenuity. I sketched out my ideas in the office when I got back: a basic 12m Warren Girder module throughout the bridge, a main span of 144m and side spans each of 84m. As with Ballachulish, a composite steel and concrete deck would be supported on sliding bearings on the cross beams, thereby separating the deck from truss actions. The north side of the river was an extensive flood plain and in order not to constrain flood flows unduly the approaches would be on an open structure of 20m spans. There was rising ground on the south side however and here the approaches could be mainly on embankment with a limited length of approach structure. This is exactly what was built some years later.

The Sungai Segama is further south, about 15km from Lahad Datu. It also had a chain-ferry crossing and although the Segama is a smaller river than the Kinabatangan it flowed through a gorge at the crossing point. A clear span of 90m would be necessary to keep the piers out of the channel and with two side spans of 60m the abutments would be well back from steeply sloping ground on either side of the river. Here my concept was for two deep steel plate girders connected by U-frame bracing acting compositely with a concrete deck. Prior to casting the concrete deck, the girders could be assembled and launched from one side across the gorge over the piers, like the Trondra and Burra bridges in Shetland, but much bigger. Again, this is exactly what was built.

Working in the office I completed the design concepts of all the bridges ensuring the arrangements at each crossing were compatible with the highway alignments and provided cost estimates of the structures to the transport economists for their cost/benefit analysis of each route. The project manager for the feasibility study was John Smith whom I knew well from Nigeria. I kept him updated on development of the bridges for his monthly reports to the Executing Agency, the Sabah JKR. We all worked hard to complete the feasibility study in the five months allocated for it by the World Bank, but we still had plenty time for leisure activities in this rather delightful tropical environment. These included snorkelling from the coral islands offshore, water skiing from the Yacht Club, squash in the KK Club and, of course, eating out in the Chinese and Malay restaurants in the town. With plenty of young engineers from Minco and KKK in the office we also formed a football team, although I think John Smith and I were the only Brits to feature in it. We won a few games against rather indifferent local opposition and then, flushed with our success, challenged the JKR. We beat them 2-0 with one of the goals something of a fluke as it was scored by me direct from a corner, sailing under the bar over the head of the goalkeeper. JKR did not take very kindly to this defeat and asked for a return which we agreed to, rather condescendingly. However, we got a shock when they took the field against us in the rematch as we could only recognise a few of the original players and the new ones looked much bigger and more determined. We were overwhelmed by this new JKR who beat us at least 4-0 and completely exhausted us in the process in 30C heat. In fact, that was the last eleven-a-side game I played in, deciding at 36 that it was time to hang up my boots. We found out later that JKR had drafted in several ringers from the Sabah State team.

When the feasibility study was completed, my wife went home with the boys, but I stayed on for another two weeks with a few others to plan for the detailed design stage to come and during those two weeks we took the opportunity to climb Mount Kinabalu over a weekend. Four of us did the climb; Mike Butterfield, Mike Dyson, Louis Akroyd who was finalising his recommendations on geotechnical aspects, and myself. We drove up to Ranau at the base of the mountain on the Friday afternoon and stayed in a log cabin at around 4,000 feet overnight, perhaps drinking rather too much Tiger Beer during the evening. The next day we climbed up through the rain forest and then on through sparse vegetation above 8,000 feet, finally reaching some huts at around 12,000 feet where we stayed until the early hours of the morning when we were due to set out again for the summit. I had a headache from the Tiger when we set off and expected it to ease with the exercise and fresh air, however, it intensified the higher we climbed and had become quite bad at the huts.

We had a guide who would lead us on the final 2,000 feet to the summit in the dark up fixed ropes on granite outcrops which were slick with water in places. It was cold with persistent rain and thunder rumbled continuously. By around 3am when we were ready to go, there was a full-blown thunderstorm in progress with lightning flashing all around and the guide was reluctant to go out saying it was too dangerous. I seemed to have been designated the leader of our group and remonstrated with him. I felt awful, having lain sleepless for a few hours in freezing cold with a pounding headache and was in no mood to listen to the guide's protestations and virtually forced him out of the hut. So up we went, all except Louis who said he felt he could not go any further (to be fair to him he was twenty years older than the rest of us). At first the rain poured down amid the thunder and lightning, but just as the darkness was lessening with dawn approaching, the rain ceased, and the sky cleared with the summit in front of us (Low's Peak, at 13,455 feet, the height later revised downwards by 8 feet after a new survey). Soon we stood at the top: to the west, the South China Sea and to the east, the Sulu Sea with the sun rising out of it. I had a bottle of whisky in my rucksack which I shared with my two companions and the guide (who brightened up immediately) and with some Germans who appeared behind us. At that point Mike Dyson revealed the reason he did not play football with us—he had only one lung—the other having collapsed as a result of some illness. Not surprisingly, he was pleased with himself for making it to the top. It was truly a stupendous view and despite my headache I was glad also that I had made the effort. It took us about an hour to get back to the hut where we had left Louis (the "Old Man of the Mountains" as Mike Butterfield called him). He made some tea for us into which I tipped the remainder of the whisky and six hours later we were back down at Ranau where my short-wheelbase Land Cruiser was waiting. My headache had miraculously disappeared as we descended below 10,000 feet. I felt pretty good if stiff all over.

There was a little sequel to the climb later that day. During the months I had been in Sabah I played squash with John Smith every week in the KK Club and as I was driving back to KK, I remembered I was due to play him that evening at 7pm. I thought I might be best to cancel the game as I was very stiff after the climb and did not think I would be able move freely about the court. However, as this was long before the days of mobile phones, I could not get in touch with him easily and decided to give it a try. I had never beaten John; he was ten years older than me, but very fit, and losing to him again seemed inevitable. Once back in the town, I picked up my kit and went to the KK Club where I got changed and walked out to the courts. These were in a separate building down a little garden path from the clubhouse. There was no air conditioning, but one did not really expect to have aircon in squash courts in those days. John arrived just after me and we started knocking up. As expected, I had some difficulty in moving to begin with, but in the hot, humid atmosphere in the court I loosened up and suddenly felt charged with energy and eager to get started with the match.

I beat John 3 games to nil, at least one of them to love, as I recall. The match was over in fifteen minutes, well within our time slot of 40 minutes. John was perhaps somewhat shocked and asked for repeat which I agreed to. The same thing happened; I won 3-0 again with another game to love. I was elated and in the bar we each put away two pints of gunner (ginger beer and soda with a dash of angostura bitters—the gunners in the KK Club were the best in the East), followed by a round or two of Tiger. As I was due to leave Sabah a day or so later, we did not have time for another match and John never got a chance for revenge. I can only put my surprise win down to the climb the days before in the thin air above 10,000 feet giving me some sort of oxygen boost when I got back to sea level. I should add that John and I were, and still are good friends. I had played squash with him in Nigeria and went fishing with him there and more recently we have fished together in Scotland. But we never mentioned that squash match again.

So, a year later, after completing the design of the Rama IX Bridge approaches it was back to Malaysia in 1981 for the detailed design of the Sabah bridges to be carried out in Minco's office in Petaling Jaya (PJ) which is about 15km from the centre of KL. PJ was in effect, a new town, having been laid out originally in British Malaya in the early 1950s as part of a rehousing programme during the CT Emergency. It also served as an overspill from parts of KL which were becoming overcrowded. PJ is divided into sections; Section 14, Section 15 and so on, and is a pleasant place with tree-lined streets and a wide variety of restaurants and *kedais*.

Minco now took the lead for the detailed design of the road with PFP in more of a supporting role, except for the bridges which I was responsible for, and for these I had a team of Minco engineers and draftsmen under me. After going through the usual approval process with JKR and the World Bank, the detailed design was given the green light to proceed in August 1981 which fitted in well with my arrival from Bangkok in early September. We found a rather nice house just off Jalan Universiti in Section 16 near the University of Malaya and Minco set up rooms for me and my team in a satellite office close by in Section 17. David and John were enrolled in the Alice Smith School, which at that time was near the *Istana* in the south of KL and I was ready to go. Except for one thing—I had insufficient money to pay the deposit on the house. To start with we had stayed in the old Merlin Hotel in Jalan Sultan Ismail in KL, and I had expected an advance would be made available to me to cover the set-up expenses. However, Minco told me no arrangements had been made, other than for monthly payments to me of the World Bank per diem allowance. These would cover school fees and rent, but not the necessary advance payments.

In those days there was no internet or even faxes. The only way to communicate with the office in London was by telex or by telephone through international operators. The latter was expensive and had to be booked in advance and the recipient of the call had to confirm availability for the time allocated by the operator. I needed to get money from the firm urgently, so I requested Minco to send a telex to London on my behalf requesting immediate transfer of 15,000 Malaysian dollars (a little under £3,000 at that time) to an account I had opened in the HSBC in PJ, to be repaid out of my per diem allowance over a period. There was no response. Another telex was sent and again there was no response. I sent a third telex, but this time, as well as requesting the money I said in the telex I would be returning to the UK if the money was not forthcoming within a week.

Then I got a call from Ganendra asking me to come around to see him. He had seen the telexes, of course, and was concerned I might leave which would have given Minco problems as well as PFP. I explained the position, making it clear that there was no way I could stay unless PFP provided me with the advance I had requested. Rather than try to bring pressure to bear on PFP in London, however, he said he had a better solution. He would lend me the 15,000 dollars interest free to cover the expenses and I would repay him in six monthly instalments as the per diems came in. I agreed immediately as I had not wanted to bring the posting to an abrupt end, even though I certainly would have done without an advance, as I had no intension of drawing on personal savings. He asked me to come back to his office later when he would have a cheque for me drawn on the Bank Negara in KL. In the afternoon, I picked up the cheque, went to the bank in KL and obtained the money, which I deposited in my HSBC account in PJ. The advance payments of rent were made the next day, leaving us with plenty cash for other expenses and general subsistence over the next few months as the per diems built up. I made monthly repayments as agreed and things went smoothly from then on. It was a clever move by Ganendra as it ensured I saw Minco in a good light and was keen to make a go of the assignment.

The next thing was to set up the design team to work under me. I was provided with CVs of staff with a structural background which was what I wanted. I picked four engineers who seemed most suitable: Inbaraj Abraham, Jeffrey Yu, Aaron Wong and Patrick Augustin. The first three were graduates with around two years' experience and the last mentioned was a newly qualified graduate with a good degree from Lancaster University. Minco were reluctant to release Patrick, however, saying he was the best of their year's intake, but after some argument I got my way. I also got several draftsmen with more promised as I needed them.

I had been careful to look for people with a good structural background, but without specific bridge experience and preconceived ideas in bridge design and analysis so that I could mould them into my way of working from the start. They certainly all lived up to my expectations and formed an excellent design team, eager to learn and pleased to be working on the Sabah project out of the main Minco office. My wife and I made a point of having them around to lunch early on which proved a big success and helped to make a happy group atmosphere. We were also assisted by one of Ganendra's sons, Torkil who was something of a computer specialist and was able to set up computer models for analysis of the Kinabatangan and Segama Bridges. The first Kinabatangan model Torkil set up was a plane frame, essentially just a two-dimensional single truss system, as had been used on Ballachulish Bridge eight years earlier. Non-uniform traffic distribution on the bridge deck results in the truss on one side being more heavily loaded than the other. A judgement then needed to be made on the most adverse loading distribution between the trusses on either side before the plane frame analysis can be run on the more heavily loaded truss. However, computing power was now becoming greater. In 1981 it was possible to run full three-dimensional frame models which could analyse non-uniform loading directly and Torkil managed to get one set up and running, allowing a check on the 2-D results. The 3-D model also enabled the effects of wind to be modelled properly.

This resulted in a more accurate analysis than had been possible with Ballachulish Bridge, but otherwise the analysis and design of the structures were very similar.

Without my experience of composite steel and concrete bridges and large steel trusses, the detailed design of the bridges on the Sabah Roads project would have taken much longer. Even so, four months into the design period I realised that despite the efforts of my team in PJ we would struggle to complete the assignment by the time I was scheduled to move on to Hong Kong. I needed some help from an engineer with experience of bridges to relieve me of the need to supervise and check everything being done. Unfortunately, at that point in time all Minco's own experienced bridge engineers were fully involved in local bridge projects and on issues which had arisen on a project in Johore which were causing them a lot of concern. I knew that PFP had recently recruited one or two bridge engineers for work on road schemes in the UK which had recently been awarded to the firm and asked if one of these engineers could come out to Malaysia for two months to help me. I got not one, but two: Colin Gill who would join the design team under me and Rudi Bradescu who would do an independent review of our work. Rudi was a Romanian around sixty who had escaped the German occupation in WW2 and eventually obtained refuge in Britain. He came from a village in Transylvania not far from Bran Castle of Count Dracula fame. David and John were fascinated by this and perhaps a little wary of him at first when we had him around to dinner, but Rudi was a delightful person and they soon warmed to him. He was also a very good engineer, and it was reassuring to have him review what I had done as I had taken on a lot of responsibility on my own. Now, with Colin and Rudi and with the hard work of the Minco graduates and draftsmen we were able to get everything pulled together and I could also get on with writing up the specification.

As a family we enjoyed Malaysia a great deal. The Alice Smith school was an excellent establishment and both boys were happy there with children's parties virtually every week and after-school activities such as football. There was a magic about Malaysia as many like us had found out and we were starting to settle into a happy expat existence there. I was learning some Malay which has stayed with me ever since. Ganendra had arranged temporary membership for us in the Lake Club which gave us somewhere to go for swimming and meals. We could drive down to

Port Dickson on the west coast in not much more than an hour, or, with a three-hour drive get over to Kuantan on the east coast at the weekend, to stay in chalets at the Merlin Hotel on the magnificent beach south of the town.

At the time there were few dual carriageway roads in Malaysia, other than in KL, but the cross-country roads were designed to British standards with generally good alignment and surfacing. In some ways driving on them was like driving through wooded areas in rural Scotland, but in other ways it was very different. Instead of ash and oak on either side there might be rubber plantations with regular lines of trees or stretches of dark primary forest seeming to press in on the road. The fifty miles between KL and Seremban which we often drove on at weekends when going to the west coast was very much like that, passing through alternating rubber plantations and primary forest. One could understand how the dark forest still present throughout Malaysia led to belief that ghosts or *hantus* were a threat if vigilance were not maintained. The road to Seremban was considered especially prone to ghostly visitations, whether from the malevolent witch-like *penanggalan*, a disembodied female head with entrails hanging below it, or worse still, a car driven by a headless corpse which caused many drivers to run off the road, often with fatal consequences. Mindful of the potential for Malay ghosts on this road we never used it in the hours of darkness.

As we had not yet formed the deep attachment to Hong Kong which came later, we viewed the move back there with mixed feelings and would have been perfectly happy to stay on in Malaysia. A week or two before we were due to leave, Ganendra called me to put a proposition to me; that I stayed on to design a bridge in KL, the so called UMNO Ramp. The ramp would provide a link from the elevated Jalan Kuching to the UMNO HQ building at ground level. To fit the ramp in, it would have to be on a tightly curved alignment and would be best constructed as a cast in situ concrete slab bridge, perhaps prestressed. Undoubtedly, it would be a tricky design and would benefit from a finite element analysis to obtain the design moments and shears which was now becoming possible in bridge design with rapidly increasing computer power. Torkil could help set this up if we acquired a suitable FE program. I was keen and Ganendra contacted Peter Fraenkel to discuss the proposition and agree financial arrangements.

A few days went by, then Peter Fraenkel contacted me to say that I was needed urgently for a particular project just starting up in Hong Kong as well as for work still to be completed on the Trunk Road bridges. There was some further discussion between Ganendra and Peter Fraenkel, but without agreement. Ganendra was not ready to give up, however, and suggested I might jump ship and stay on with Minco. I thought about his proposal seriously and discussed it with my wife. On one hand we were now settled in Malaysia and knew the hassle that would await us in Hong Kong; starting the boys in a new school when they were happy where they were and finding suitable and affordable accommodation in Hong Kong which would not be easy as I knew rents were high. I was sure I could negotiate with Ganendra to get matching or better terms and conditions than I was currently getting with PFP. But on the other hand, it was a step into the unknown and while there was plenty work in Malaysia at the time, there was no guarantee that would continue. PFP were likely to get other work in the Far East once the current jobs came to an end and if not, there was always the UK offices. We decided that Hong Kong it would be, therefore.

We packed up and said our goodbyes to the friends we had made. My design team all came to the airport at Subang to see us off. I was sorry to leave them as they were all very pleasant young men and had worked hard on the design of the bridges. In fact, Patrick and Inbaraj eventually started a consultancy of their own with whom PFP associated on several projects in Malaysia some years later and we have remained friends to this day. Rather than fly direct to Hong Kong we had arranged to take a ten-day holiday in Penang. I had been there briefly on a promotional visit to the Port Authority in Georgetown after which I had taken a taxi to a hotel at Batu Ferringhi for lunch. I liked the beach which was still relatively unspoiled at that time and that was where we headed to stay at an hotel called the Casuarina which had just opened. For the ten days we lazed on the beach, hired a small fishing boat for trips to Monkey Beach further along the coast and enjoyed Penang food in the evenings. Then it was on to Hong Kong.

Eastern Suns and the Autumn Moon

> That is the land of lost content,
> I see it shining plain,
> The happy highways where I went,
> And cannot come again.
> *From A Shropshire Lad, A. E. Housman*

On completion of the design of the Sabah bridges, we arrived in Hong Kong from Penang on 7 April 1982 and checked into adjoining rooms in the Royal Garden Hotel which had just opened. It had been constructed on reclaimed land on the Harbour side of Chatham Road in the area which came to be known as Tsim Sha Tsui East. We were on a high floor in the hotel, and I remember looking out of the window on the evening of our arrival as the sun was setting in the west through a haze over the urban landscape of Kowloon. Somehow, it was not an encouraging sight after the green of Malaysia.

That first week in April had seen the start of the war between Britain and Argentina, following the Argentine landings on the Falklands. The South China Morning Post was full of the preparation of an invasion fleet to sail from Britain to the South Atlantic to retake the territory.

Our priority was to find accommodation, as a residential address was needed to get children into one of the ESF schools. This had to be largely down to my wife as I was immediately plunged into work in the office. Her search ranged widely across Hong Kong from Mid-Levels on Hong Kong-side to Kowloon Tong and the New Territories. It soon became clear that my rent allowance would not stretch to a flat in any of the better urban areas and she focused on the NT. She was greatly aided by a pleasant Chinese girl from an estate agent firm who helped her zero in on several possibilities near Sai Kung and along the Clearwater Bay Road, areas we knew well from 1977. The place we particularly liked was around a mile and a half north-east of Sai Kung in a little settlement called Long Keng. The house had four bedrooms, a big lounge cum dining area, a veranda with a view over old paddies to the hills and a garden beside the house. Most importantly, we could just about afford the rent. However, the owner was still putting the final touches to it for letting and it would not be ready for two months. Then we had a brainwave: we would rent temporary accommodation nearby until the house was ready. This would get the boys into school—in the Sai Kung area the ESF school would be Boundary Junior School in Kowloon. As it happened, there was a village house for rent through the same agency in Tai Wan Village about halfway between Sai Kung and Long Keng. Ideally, the owner would have liked a minimum six-month lease, but he settled for two months and just two weeks after our arrival in Hong Kong we moved in. At the same time, we negotiated with the landlord of the house in Long Keng and put down a deposit to secure it when ready. Entry to Boundary Junior for the boys was quickly arranged using the Tai Wan Village address. To round all this off, I bought a second-hand car and picked it up the day we left the hotel.

We were pleased to have sorted out accommodation so quickly, largely down to my wife and the help she had from the girl in the agency. We were also lucky in the landlords themselves. The house in Tai Wan Village was owned by Brian Huddleston, who had taken it on with a friend in the Football Club of which he was a member (Brian was a good rugby player and had represented Hong Kong). His friend had then reneged on the deal leaving Brian to pay a mortgage on the house on his own. Brian had a Chinese girlfriend, Carol, with whom he stayed in her flat on Hong Kong-side and even two months' rent would help with his mortgage. He was easy going and we became friendly with him and often saw him until eventually he left Hong Kong for a job in the United States.

There was one drawback to Brian's house—despite being on a steep hillside it was prone to flooding in heavy rain which we found out shortly after moving in. It was surrounded by traditional Chinese village houses in terraces cut into the hillside above the road through Tai Wan Village. There was a stream on one side of the house which normally had just a trickle of water in it, but which became a raging torrent in heavy rain. In the stream just below the house there was a culvert protected by a screen, intended to trap the rubbish thrown into the stream bed by the villagers and prevent the culvert blocking up. Unfortunately, so much rubbish accumulated in the dry winter season that the screen itself could become blocked causing water to back up and flood into Brian's house. This happened on several occasions in a short space of time in May 1982 when a slow-moving area of low pressure sat over Hong Kong for three days, depositing nearly 600mm of rain on the colony. On each occasion there was nothing for it but to mop out the water and silt making sure Brian and Carol were summoned from Hong Kong-side to help (and to safeguard Brian's furniture which Margo had managed to raise above water level). We were relieved when the house at Long Keng was pronounced ready for occupancy.

The owner of the house was Mr Wong Wai-Cheung whose main occupation was pig breeding at Tso Weng Hang on the Tsai Mong Sai Road about half a mile away. He turned out to be an excellent landlord and helped us get settled into the house at the start of the lease. He also found us a part-time amah, Ah Yeung, whose husband looked after a government water pumping station nearby. The house was on Lot 767 DD 216 (the DD in the postal address refers to District for Development) and was a typical modern village house as seen all over the NT in Hong Kong. These were products of a special arrangement the colonial government had come to with residents of the NT known as the Small House Policy. This entitled every indigenous male descended through a male line from a resident of the NT in 1898, to obtain a grant to build a house of not more than three storeys on a plot of 700 square feet. It was promulgated by David Akers-Jones in 1972 as a trade-off to ensure the support of residents of the NT for establishment of the new towns. We moved into the house in Long Keng in June, with Mr Wong still finishing some things off. British forces had just retaken the Falklands with the final assault on Port Stanley and Mr Wong was as pleased as we were with the news.

The downside of Long Keng was its distance from Kowloon which meant lengthy journeys to school and work. There was an efficient school bus service, however, called "Passengers", which had been set up by an enterprising English lady for the many children in the area attending the ESF schools. Every morning at 7.30am the boys were picked up by one of the buses at a roundabout near us on the Tsai Mong Sai Road. For my part, Margo would run me to Sai Kung where I would get a light bus to Choi Hung and then go on from there by MTR to Jordan station near the office. A Passengers bus would return the boys to the roundabout after school, and I would make the morning's journey in reverse. One soon got used to this routine which remained largely unchanged for some time until I joined the Kowloon Cricket Club and could park at the club. Then, with family membership of the KCC, there were definite advantages to living out in the NT. At weekends we could drive into Kowloon to the club on relatively traffic free roads while day trippers to the Sai Kung Country Park filled up the roads heading in the other direction; and in the evening when we were going home, our road out of town would be quiet with heavy traffic coming the other way. But I suppose the main thing about Long Keng was the fresh air and the views from our veranda across the abandoned paddies to the hills and mountains of Kowloon, rising to Ma On Shan (Saddle Peak in Cantonese) in the west, at 700 metres, one of the highest peaks in Kowloon. Sitting on the veranda with a beer in summer while the barbecue heated up and watching the sun set behind Ma On Shan, more than compensated for the long journeys.

With domestic arrangements sorted, I settled into the office quickly. I knew most of the staff, both British and Chinese, particularly Tony Bowley and Ted Foster from various periods in Hong Kong between work in Bangkok and Malaysia. Ted was a lot more proactive than his predecessor and had insisted in being made a partner of the business in Hong Kong, as well as negotiating other provisions such as a company car. His wife Joan remained in Australia, but she came over to Hong Kong for a few months several times a year. He lived quite close to us near Sai Kung, and we saw a lot of him and Joan when she was in Hong Kong.

Lamma Island Cement Terminal

When I arrived, although the Trunk Road was still the mainstay of work in the office, the works were only just gearing up and initially I was involved in the construction of a cement terminal at Sok Kwu Wan on Lamma Island. The client was Far East Cement Company, a subsidiary of Shui On group who planned to store cement on the terminal for distribution around Hong Kong on barges. I had no prior involvement in this job as I was away from Hong Kong when the design was carried out. Some of the features seemed to me not well thought out, but as a contract had already been awarded to McConnell Dowell, a contractor from New Zealand, there was nothing to be done other than get on with it. The terminal was designed for cement carrier vessels of 20,000 tonnes. It had a jetty head on which a cement unloader was located, along with berthing and mooring dolphins and a separate berth for the barges. Onshore, there were two cement silos with conveyors and bucket elevator systems to feed cement from the unloader into the silos. The silos each had a capacity of 20,000 tonne of cement and were 24 metres in diameter and 40 metres high with walls 0.3 metres thick. The design of the silos was not fully detailed, and I was landed with the job of completing it.

Miro Huq, who had been in Kano with me, had been brought out from the UK on a short-term basis to be RE on the contract. The first engineering problem we encountered arose with piling for the jetty head and dolphins. Steel tubular piles 0.6 metres in diameter were used to support these structures of which a lot were raking piles designed to resist lateral forces from berthing the vessels. Where these raking piles were in tension there was insufficient pull-out capacity in the embedded length of piles to resist the tension and ground anchors were to be installed through the piles and drilled and grouted well into sound granite rock below the superficial clay soils of the seabed. The piling operation had just got underway when I arrived, and the piles were in the process of being driven through marine clays to the rockhead. Drilling rigs were set up for the ground anchors, but it was found that the drill bits could not get through some of the piles, hitting obstructions of some sort in the shafts. It had been expected that the shafts would plug up with clay on driving the piles, but clay plugs should not have offered any resistance to the drill strings. Fortunately, the pile caps on the marine structures had not yet been cast and one or two of the blocked piles could be extracted without much difficulty. Inspection showed the piles were distorted at their lower ends which was what was preventing the drill bits passing through.

What had happened was the piles had hit core stones within the decomposed granite underlying the marine clay some way above sound rock and the open ends of the piles had buckled. When weathering of rock is uneven some of the original hard granite can be left as boulders above the main rock horizon. The question was how to get around the problem without a complete redesign with all the financial implications that would entail. I decided to try a simple modification to the piles. I had the buckled ends cut off and replaced by welding on lengths of high tensile steel tube of the same diameter with a shaped cutting-edge which I hoped would split the core stones it encountered. This was first tried with the two piles which had been withdrawn and it worked. All the other damaged piles were then pulled out, modified in the same way and re-driven without any problems, allowing the anchoring process to be completed. There was no science in my solution, just seat of the pants engineering, but I used high tensile shoes on tubular steel piles ever since.

The silos had identical concrete domed roofs 100mm thick, and the question arose in my mind as to whether this thickness was adequate. There seemed to be no design calculations or indication how, or by whom, the thickness had been determined. There could be no snow loading of course, but as well as its own weight, each dome had to resist point loads spaced around the crown of the dome from the framework of the bucket elevator system which fed cement into the silos through an aperture. I obtained a textbook on classical dome theory which showed that 100mm thickness was more than sufficient for self-weight of the structure, but it gave little guidance on point loading effects. All I could do was to convert the dome into equivalent arch strips with a rise and span the same as the dome and analyse the arches for stability under the point loads. This required an assumption of the effective width of the arches, but otherwise must have been conservative as it ignored the 3-D action of domes which made them such efficient structures for cathedral roofs. My crude analysis showed 100mm to be adequate and I left it at that.

As it happened, I was able to test the capacity of the domes at first hand. McConnell Dowell had decided to construct the silo walls by slip-forming with moving shutters raised in continuous operations at a uniform rate as concrete in the forms sets. It is a common method of constructing silos and towers of regular plan shape. Wet concrete is fed into the forms at the top and with the right speed of raising, hardened concrete emerges below the shutters at the bottom. An Australian contractor specialising in the method was engaged as subcontractor to McConnell Dowell and they came up with an innovative method for constructing the silos. Once the foundations for each silo had been completed the domed roof was cast on its own shuttering on top of the foundation. When the concrete had set, the completed dome was connected to special steel rods within the reinforcement cages for the silo walls by through-jacks which could climb up the rods above the moving shutters ahead of the slip-form system taking the dome up with them. The rods took all the weight of the dome together with various items of construction equipment and bundles of reinforcing bars needed to form the steel cage for reinforcement of the silo walls. By this means a self-contained working platform was provided which moved upwards with the slip-forms. The loading on the dome in this process would be at least equal to the permanent loading it would receive in service, a good test for its structural capacity. Construction of each silo was estimated to take five days with a continuous rate of climb of around 0.3m an hour. As this would be a 24-hour round-the-clock operation Miro had to have help with supervision and we put a rota system in place. For my part, I did one overnight supervision slot for each silo.

There were two structural issues to be considered. First, the steel rods within the forms which supported the dome were only 30mm in diameter and each rod carried several tonnes of dome. If the jacks which clamped on the rods were too far above the hardening concrete in the forms, the rods would buckle and the whole assembly would collapse. I worked with the sub-contractor to agree the maximum safe distance above the concrete for the jacks as these had to be raised on the rods periodically as the slip-forms moved up below them. Clearly there was an advantage in raising the jacks in as large increments as possible so that they did not need constant adjustment, but I had no problem in persuading the contractor to abide by the 1.0m limit I set—we would be on the dome during the slip-forming operation and our lives would depend on the limit being adhered to.

The second issue was the loading from reinforcing steel bundles on the dome prior to fixing these into cages in the silo walls. A tower crane located beside the silo placed the bundles on the dome as it went up and lifted skips of wet concrete for the forms. I did not want loading on the dome to become to be too concentrated at any point, simply because I was not able to analysis the effects of the loads with any accuracy. To get around this I stipulated a maximum weight of reinforcing steel on the dome at any time which had to be distributed around the dome as evenly as possible to approximate to a uniformly distributed loading which I could analyse by classical dome theory.

It all worked well, and I enjoyed my two nights on the silos, although with little to do other than make periodic checks on concrete quality and the jacking arrangements things were rather boring, especially around 3 or 4 in the morning when one's eyes were getting heavy. It was August when rain and thunderstorms might be expected, but fortunately both silos were completed in dry weather. The Australians brought equipment and personnel (they worked 12-hour shifts on the silos) to the site on a small tug from a jetty in Sandy Bay on the west side of Hong

Kong Island. On each occasion I was on site overnight I left my car in Sandy Bay near the jetty and came over on the tug with the night crew in the early evening, to return with them in the morning on completion of their shift. Being Australians, the first thing they did on the return journey was to break out supplies of San Miguel. After a night's supervision, a couple of cans went down rather well even at eight in the morning.

There was one other issue on this project which is worth recording. This related to the foundations of the silos. The silos were located on reclamation below an old quarry where granite had been mined for use as filling and concrete aggregate. The flat area below the cut slopes had been extended by filling into the sea to give a larger working area at the quarry. Heavy structures such as cement silos would normally be supported on piles, but in our case, there was, quite rightly, a concern that piles would encounter boulders in the reclamation which had probably not been carefully controlled, and the silo foundations were therefore in the form of deep concrete rafts. A site investigation had shown up to 10m of fill over at least 5m of marine clay which had formed the original seabed. This clay should have largely consolidated since the reclamation had been there for a considerable time. But as the two silos were closely spaced, the ground pressure bulbs under each foundation raft would intersect at the level of the underlying clay and it was expected this concentration of ground pressure would cause some further consolidation. Geotechnical calculations showed there would be a tilting of the rafts towards each other by up to 150mm and allowance for this movement had been made in the design of the bucket elevator systems on the sides of the silos.

To check that the settlement of the silos was as predicted, a level survey of their foundations was carried out after their completion and every three months during the maintenance period. The first two surveys indicated virtually no settlement of any kind, but once the silos were filled with cement and came into full operation, it was a different story. At the nine-month survey definite settlement was detected, but the settlement profile was in the opposite direction to what had been predicted—the silos were tilting away from each other. At first, I thought the surveyor had mixed up his easts and wests, or perhaps the accuracy of the survey was not what it should have been, and I expected things to have sorted themselves out by twelve months. But at twelve months the outward tilts were even more pronounced. Had we stumbled on some hitherto unknown phenomenon of soil behaviour? The geotechnical engineers could offer no logical explanation, but I knew that if this tilting continued there would soon be a problem with the operation of the bucket elevators.

The cause was discovered in routine inspection of the mechanisms for discharging stored cement in each silo to a central low-level conveyor between them which transported it to the barge berth. These were what were termed air-slides. They introduced air into the stored cement which made it behave like a liquid and flow into a chute leading to the conveyor. There was an air-slide and a discharge chute in each silo. These faced each other on either side of the central conveyor. In the inspection it was seen the cement did not settle evenly. It drew down unsymmetrically above the air slides leaving a wedge-shaped heap of cement against the outer walls of the silos opposite the chutes. This was producing a markedly non-uniform loading on the rafts, hence causing the outward tilts. The air slides had been supplied by Mitsubishi Heavy Industries and a representative of the company was brought in from Japan. He agreed that it was down to Mitsubishi to fix the problem as the air-slides were not working as advertised. They were modified to inject higher volumes of air into the cement mass aiming to draw down the cement more evenly to equalise the loading on the rafts. This was successful and the tilting eventually stabilised just within the tolerances of the bucket elevator systems.

Trunk Road Construction

Once the Lamma Island terminal was near to completion, I was able to focus more fully on the Trunk Road construction which had now started in earnest under two overlapping contracts: an advance contract for a section of road and bridges at the north end of the route and the main contract comprising 7.3km of dual 3-lane carriageways in the south. This main part was constructed largely on reclaimed marine embankment. The advance contract was undertaken by Leighton Contractors Pty and the main contract by Vianini Lavori SpA. For me, there was some bridge design still to finalise. There were also ferry terminals in Tolo Harbour to be relocated clear of the road. I was to be responsible for Head Office responses on all structural issues which might arise during construction. It was an advantage that I knew the senior site staff well. John Tainsh was the CRE with Martin Broadgate his deputy, as in Shetland. This was now the third time Martin and I had worked together on the same job. Willie Roxburgh from the Glasgow office was also with us in the office, and he worked on the bridge design as well as dealing with one of the ferry terminals. He also distinguished himself by representing Hong Kong at rugby in regional games and at the Hong Kong Sevens.

At the start of the main contract 3.1 million cubic metres of soft deposits were dredged along the route by a combination of chain bucket, grab and suction dredgers. The submersible filling was obtained from weathered granite outcrops in the Whitehead borrow area on the western slopes of Ma On Shan on the east side of Tolo Harbour. This was brought to the site on derrick barges and split-hopper barges from temporary load-out jetties. Great care was needed in initial filling operations along the route to prevent the formation of mud waves in the 2 to 3 metres of soft deposits which had been left in place. When the submersible filling reached a level of +2m above PD, 2 million cubic metres of compactible fill was placed to bring the embankment up to the formation level of the highway. The compactible fill was obtained from three borrow areas on the hillside on the landward side of the KCR. Haul roads and temporary bridges over the railway had to be constructed at some cost, but the operations were kept off all public roads so that the contractor could use vehicles which were not road-licensed, running on cheap industrial gas oil. The saving in the cost of fuel over the course of filling more than offset the expenditure on haul roads.

In 14 of the 16 bridges in the scheme, driven piles were used for support of the piers and abutments. These were either 305 x 305mm steel H-piles, or 508mm diameter steel tubes. There were 1550 piles in total. The piles were driven to "refusal" in volcanic rock by diesel piling hammers and at least one pile at each foundation was tested to 1.5 times its working load. A small footbridge over the KCR near the Chinese University was founded directly on volcanic rock outcrops. The single span prestressed concrete footbridge (by now named "Treatment Works Bridge" after Sha Tin sewage treatment facility which was close by) was founded on 1.0m diameter hand-dug caissons.

The use of hand-dug caissons in Hong Kong has long died out, but they had a lot to commend them. They were a traditional Hong Kong method of forming piles, often carried out as a family enterprise with the husband excavating the shaft and casting successive thin concrete rings to keep it open, while his wife and often children helped him and disposed of spoil from the shaft brought up in buckets with simple lifting tackle. The shafts at Treatment Works Bridge went down 20m in places and I inspected the bases of most of them as I had done with the bored piles in Glasgow at the start of my career. Once the work was accepted, a cage of reinforcement would be lowered into the shaft which would be filled with concrete of the specified strength. This low-tech, but highly effective method of forming piles was widely used throughout Hong Kong at that time, but it would now be considered taboo on health and safety grounds, although as far as I know accidents were rare. I chose hand-dug caissons for Treatment Works Bridge because one end of the bridge was close to the KCR which was on a curved embankment at that location and did not want vibration from driven piling causing settlement of the tracks. I was later praised for the wisdom of my choice for another reason. A lot of boulders were encountered in sinking the shafts which would have impeded driven piles, but these were easily broken up by jackhammers in the hand-digging operations. In fact, prior to forming the caissons, I knew nothing of the presence of the boulders which had not been revealed in site investigations, but that did not stop me from accepting the praise for my wise decision.

However, there was one issue in the construction of Treatment Works Bridge which caused me acute anxiety at the time. Again, this was related to the proximity of the KCR. In order to create space for the approach ramp to the bridge on the west side we needed to excavate into part of the railway embankment and there was insufficient room for a conventional retaining wall. I opted for a stressed, rock-anchored structure instead, reasoning that the ground compression from stressing would restore any loss of stability as a result of the excavation. The main structural components were precast concrete "soldier" beams spaced 3m apart in slots cut into the embankment, through which cables inclined downwards at 45 degrees were grouted into rock below the railway and then stressed to induce the ground compression. Horizontally spanning concrete slabs were then cast between the soldiers to retain the embankment soil once it had been trimmed to the desired profile. Construction was tricky to say the least. In one area the embankment consisted of a particularly loose sand and material from it trickled down steadily as we fitted in the soldiers, especially when trains thundered past only a few metres above us. We had agreed with Vianini to excavate for only one soldier at a time, then to drill and stress the anchor before repeating the process with the next soldier. The gaps between them were not to be excavated until soldiers were in place and stressed on each side of the gap. I got a call at home one Saturday afternoon from George Anderson, a Shetlander and a particular friend of mine who was the RE for the structures. A lot of sand was coming down and he was afraid the embankment might collapse. Driving over to the site that afternoon, I tuned in to RTV Hong Kong, the official media channel like the BBC in the UK, fearing to hear that a collapse had occurred with a train on the curve. When I got there, I found that the soldiers on either side of the problem area had been stressed up which was a relief, although some sand was still coming down. George and I agreed to instruct Vianini not to excavate any further till all the soldiers around the curve were in and fully stressed up, before excavating and casting the infill slabs one at a time in a hit and miss pattern. I was back for much of the Sunday with George and every day during the following week while this was being done. Once everything was stressed and the infill slabs were cast, I was confident the embankment was completely stabilised. The finished job, to my eye at least, looked pretty good without any hint of the rather fraught time we had in its construction.

At the peak of the bridge construction, I spent a lot of time on site, often in the site office compound with John Tainsh, Martin Broadgate or George Anderson as there were always many minor issues relating to bridges to be dealt with to keep the work flowing. PFP had a site establishment of around fifty on supervision and Vianini had about the same number on the works in addition to numerous subcontractors who are always part of the construction scene In Hong Kong. The office compound had a sizeable dog population, and it was noticeable that the number of puppies around in winter decreased markedly at Chinese New Year. Barbecued dog was still a popular dish amongst the Chinese site staff despite it being illegal to eat dog in Hong Kong.

There were over 400 precast concrete post-tensioned beams for the bridge decks which were cast at a central yard as I had intended. This was adjacent to the main concrete batching plant on the site served by a travelling gantry crane. The concrete in the beams was a high strength for the time (Grade 45/20 and 52.5/20 for the longest beams) with an agreed plasticiser in the mix to improve the flow characteristics of the concrete. Steel formwork was used to give maximum reuse with both internal and external vibrators. It was hoped these provisions would ensure a high quality of workmanship in the beams. The side forms were struck after 24 hours in winter and 16 hours in summer and the beams were generally stressed at 4 or 5 days depending on strength gain in trial concrete cubes.

When the very first beam had been cast, I inspected it with John Tainsh. I was appalled to see it was badly honeycombed and generally poorly formed with broken arises and other defects. My concerns on using precast beams made from Hong Kong concrete immediately resurfaced. John was not at all perturbed, however, and merely told Pitassi, a likeable Italian who was Vianini's 2ic on the contract, that the beam could not be used in the works. Pitassi protested that it could be repaired in his voluble Italian way, but John was adamant. Eventually, Pitassi accepted this and said he would have it taken away and broken up, but John would only let it be moved clear of the bed and instructed Vianini to leave it in the yard. There it lay for the duration of the works, *pour encourager les autres* as John said. This certainly had the desired effect as all other beams for the works were cast to the highest standards. It also demonstrated the value of a good, old fashioned Engineer's Representative, prepared to take responsibility for the works in the days when we had a sensible form of contract without the plethora of project managers and advisers we have now.

There were no hard feelings from Pitassi on the issue of the condemned beam and I generally got on with him very well. He would often appear to make a drama out of impositions we put on Vianini for the good of the project, but it was all an act, and he invariably had things done properly with some droll comments thrown in. He got married while the contract was underway to a very pretty girl from Uruguay. Most of the PFP site staff along with Tony Bowley and I, and our wives, were invited to the wedding. The reception was held in the grand surroundings of the banqueting hall of the Peninsula Hotel. After some drinks, we were shown to our places for the meal, and I thought there must have been some mistake as we were taken to the top table. But no, we were assured, this was intended as we were honoured guests. Margo and I were placed next to Pitassi's father, very much an Italian gentleman in the nicest possible way, who was the best of company. It was his first time in Hong

Kong, and he kept us amused with his observations and impressions of Hong Kong life as it appeared to him.

In addition to the bridges, other structural works in the project included culverts (one of four cells in the reclamation at Pak Shek Kok), two ferry piers, a public pier and a vertical seawall. I was also responsible for a sewage pumping station and a rising main at the Chinese University, working with a freelance Greek engineer called Paul Trantallis whom we engaged to design the electrical and mechanical elements in the facility.

The main works were completed in 1985 along with a section of the NT road system immediately to the north. The road was officially declared open by the governor, Sir Edward Youde GCMG on 24 September 1985, and to the public the next day. A charity walk along the road was held that day raising HK$1,000,000 for the Hong Kong Community Chest.

Looking back on it, the project was a major achievement for PFP to have taken on and complete successfully virtually from a standing start in the colony and it should have acted as a springboard for even bigger things. This did not happen despite the firm having made substantial profits from the Trunk Road which could have been reinvested to increase the firm's profile. Ted Foster got a small percentage of the profits through the partnership holding he had negotiated when he took up the appointment, Tony Bowley and I got bonuses, but nearly all the profits went back to London and no appreciable investment was ploughed back into the business in Hong Kong. It could be argued the partners had risked their own money in setting up the office back in 1974 and pay-back time had arrived. But this was short-sighted. Large infrastructure projects in Hong Kong were on the horizon. Other British consultants recognised this and geared up for them accordingly. The opportunity to become the major player envisaged by the PWD was not grasped. It was a pity to have covered the hard yards and not pressed on further, but not even Ted challenged the policy and PFP soon dropped to second division status as large London firms set up offices and built up in their Hong Kong operations.

However, for me there was still interesting work to come in Hong Kong through the remainder of the 1980s, and in Malaysia and Thailand where the Sabah Roads and the Rama IX Bridge were now under construction. Also, I was very happy in Hong Kong. Our third son, Stephen, was born in 1983 in the British Military Hospital in Wylie Road in Kowloon. The older two boys were settled in Boundary Junior School and playing mini rugby with Kowloon Rugby Club. We also had family membership in the KCC with all the facilities the club could offer. This duality of work and play in Hong Kong was one of the things that made it such an appealing place for expatriates.

Site Formation at Kai Liu

For some time, Ted Foster had been chasing the Hong Kong Housing Authority (HKHA) for consultancy commissions. He had got to know Dennis McNichol who was the Assistant Engineering Director of the executive arm of the Authority, the Housing Department, and had taken him to lunch at the Hong Kong Club, something always appreciated by government engineers in those days. I had never really considered HKHA as a potential client, thinking of them only in terms of building the vast public housing estates springing up all over Hong Kong at that time. In fact, the preparatory civil engineering work needed before the estates could be built, often exceeded the cost of the housing itself and was of vital importance if the ambitious government housing programmes in the 1980s were to be met. Civil engineering was becoming an even more critical factor as flat land for housing was running out and new estates were now mostly being built on the steep, unstable slopes of Hong Kong Island and Kowloon. It took about a year, but eventually Dennis succumbed to Ted Foster's overtures and said he might have a job for us. I accompanied Ted to a meeting in the HD offices in Ho Man Tin to discuss this.

Dennis outlined the importance of the civil engineering site formation to HKHA developments. This covered everything that had to done to make a site ready for housing construction. On the hillside sites now being developed it would normally include stabilising slopes, forming building platforms, constructing access roads and providing drainage. On completion of the civil engineering contract, the site had to be completely ready for commencement of the building contract. Developments were planned in their various phases to overall completion at a target date for estate occupancy some years in the future, and the first phase was site formation. Any delay in the first phase could carry through to delay the final completion with a loss of rent for the delay period. Where the delay was caused by the contractor, this would form the basis for calculation of liquidated damages in the contract.

What Dennis had in mind for us to start with was a relatively small site formation project in East Kowloon near Kwun Tong. The development was called Kai Liu and was close to the site of an old village of that name which had been demolished for earlier public housing in the 1950s. There was a steep rocky hillside to the west with squatter huts on it rising to a level area on which the Bishop Hall School had been built. To the north there was a controlled refuse tip and to the east the Tsuen Kwan O Road (later upgraded to form the approach to the Tsuen Kwan O Tunnel). There was also a water course running down from Sau Mau Ping in the form of an energy-reducing cascade feeding a stream in a nullah on the level ground below. The housing development to follow in the second phase was to consist of three "Trident Blocks", each of 36 storeys at the base of the rock slope. Our job would be to have the slope stabilised and cut back to make room for the blocks and the nullah culverted to suit. Access roads would be required with the controlled tip reshaped and its leachate which simply ran into the storm drainage system of the area intercepted and piped off the site. The programme for design and construction of the site formation works was to be 21 months.

Of course, we indicated we would be delighted to take on the job and Dennis said he would be happy to engage us with just one condition; we would need to engage a special geotechnical adviser for the rock slope. This was Len Threadgold who had done similar work for HD in the past. Len was based in England where he owned a small geotechnical consultancy, but he would visit Hong Kong as necessary to fit in with our schedule. There was no fee negotiation as in those days; standard government fee conditions applied to all civil engineering consultancy work. These conditions were not overly generous, but they were adequate and ensured consultants could operate profitably which was in everyone's interest, something which has been lost sight of in the days of unfettered fee competition. Dennis then introduced us to Steve Riley, a senior engineer, who would be our main contact in the HD, along with an engineer, K. K. Mak (Peter) for day-to-day matters. Thus, began a fruitful relationship with HKHA.

Kai Liu means "Chicken Coops" in Cantonese and at the time we were engaged in the project I just accepted that was what the site was called and only later tried to find out where the name came from. It seemed there had been a village of that name on the site which had been cleared in the early 1950s for one of the very first public housing developments in Hong Kong. This development was located further down the nullah and consisted of several of the "H-Blocks" used for early public housing schemes. These were buildings of five storeys, H-shaped in the plan, with accommodation units of 180 square feet on each floor off long corridors in the legs of the H. It seems likely that the inhabitants of the village which was cleared had been relocated to the H-Blocks. But that does not explain where the name came from in the first place. It was unlikely with such a name it was a normal village and perhaps it was a derogatory name coined by long-term residents of the area for a collection of cramped huts thrown up by the first squatters. That would explain why it had been cleared as the government would never have removed a long-established village of Hong Kong residents with property rights, no matter how pressing the housing need. In any event the name Kai Liu has now disappeared from Hong Kong as the public housing estate built on the formation which we created was given the name Tsui Ping South Estate.

I had total responsibility for the project and for the first time acted as Engineer on a contract when construction got underway. Prior to that we proceeded with a site investigation. Its main aim was to determine the degree of weathering and jointing of the granite rocks in the slope to enable it to be steepened in a stable manner. Len provided excellent advice. He introduced me to rock mechanics and showed me how to determine of the stability of cuttings in a rock mass from the pattern of joints within the mass. Different mechanisms of failure can arise in rock cuttings depending on the orientation of the cut with respect to the jointing system; the main ones being toppling of slabs of rock, sliding of blocks and failures of wedges where joints intersect. The orientation of rock joints in the slope at Kai Liu was obtained from inspection of rock outcrops and from these, Len constructed "polar diagrams" representing the predominate joints in the mass. Potential instability and failure from one or more of the mechanisms could arise where the joints "daylight" on the cut faces. By this means we were able to design the profile of the cut slope to avoid daylighting joints and any risk of instability. Where a lot of water is present in parts of slopes, horizontal drains can be installed to reduce water pressure in the joints which would further destabilise them. The design profile for the steepened slope at Kai Liu was a series of berms or terraces with horizontal drains in wet areas. Similar slope treatments can be seen all over Hong Kong.

The platform for the Bishop Hall School at the top of the slope presented a particular problem. This had been formed by the PWD in the 1960s and in order to trim back the rock below the platform the fill slope in front of the school would also have to be cut back, but it was already at a steep slope close to its limit of stability. A conventional retaining wall was a possibility, but this would have required cutting into the fill slope to form a base for the wall which would have endangered the slope until the retaining wall could be built. I decided on a two-stage approach. First construction of a vertical wall in front of the school, formed by hand-dug caissons taken down through the fill of the platform without disturbing it and embedded in rock below. Then removal the filling in front of the caissons which would act as vertical cantilevers to retain the platform.

On the Trunk Road, the NTDD had left us to get on with the design of the road and bridges with little technical involvement in what we were doing, other than keeping them up to date on progress through their engineer Rob Dickie (actually, Colonel Rob Dickie, on secondment from the Corps of Engineers in the New Zealand army). However, for the slope works at Kai Liu we had to satisfy the Geotechnical Control Office that our design was safe with no likelihood of failure. This was especially the case at Kai Liu as the site was close to Sau Mau Ping where previous slope failures had led to considerable loss of life. On two separate occasions, poorly compacted fill slopes had suffered mud slides in heavy rain and had engulfed the lower storeys of public housing, burying many people alive. After the second failure a technical report on the mud slides had been prepared by Binnie & Partners and a public inquiry had been held. The GCO owed its existence to the findings of the inquiry as one of its recommendations had been to establish a special geotechnical office within the PWD with the technical capability to check and monitor existing slopes all over Hong Kong. New slopes were also to be subject to the approval of this office. I discussed the design of the caisson wall with the GCO and particularly how close together the caissons should be sunk to be fully effective in a retaining function. Based on experience elsewhere in Hong Kong the GCO advised that the caissons should be not more than one diameter apart. The basic trimming of the rock slope was also discussed and with Len Threadgold's help on the rock cutting our complete slope design was approved by the GCO without any major amendment.

During the design period arrangements had to be made to clear the hillside of the squatter huts. These housed some of the very poorest people in Hong Kong, nearly all of them having come in from the mainland. Generous compensation was provided by the government with rehousing of the squatters in public estates as close to Kai Liu as possible to avoid disruption to their daily lives. I spent quite a bit of time on that hillside during the site investigation, threading my way along paths between the huts linked by festoons of electricity cables which tapped illegally into the CLP grid. There was no animosity from the squatters at being uprooted from their huts as they were pleased to be moving to what they considered would be better housing and the compensation for the move was also appreciated. I was struck by the efforts they had made to keep the hillside tidy with little gardens around the huts (unlike the residents of Tai Wan Village in the NT), and by the immaculate uniforms of the many children living there when I saw them making their way to and from school.

Once the design had been completed, we put the job out to tender. The lowest bid came from a Korean contractor, Korea Shipbuilding and Engineering Company, who were new to Hong Kong. The price was over 15% below the next lowest which gave some concern. The tender assessment we carried out did not indicate anything untoward in KSEC's pricing; their rates were just lower than the others across the board. Often a low bid like that would be struck out as under-pricing could result in financial pressure on a contractor leading to poor workmanship and constant claims and a risk of the contractor going bust. I hedged my bets in the tender assessment report, without a giving a definite recommendation, leaving the final decision to the HD and after some discussion with Dennis and Steve it was decided to accept KSEC's tender. As most of the structural work on the Trunk Road had been completed by this time, George Anderson was available, and I brought him in as Engineer's Representative with two inspectors under him. He was glad to take up the appointment as he had not wanted to leave Hong Kong. We were ready to go.

In fact, KSEC performed quite well, although they sub-let a lot of the work content to local Chinese contractors as is usual in Hong Kong. I liked their site agent, Mr Lee who worked very hard to satisfy our requirements. The main problem, however, was a cultural one. Their attitude to safety was casual to say the least and

I was constantly issuing warning letters in that regard. Under the Hong Kong Government Conditions (as with the ICE Conditions), the Engineer had the power to temporarily suspend work on a contract if in his opinion the work was not being carried out safely and I did this on one occasion and threatened it on others. This power was not often used, but the time I did so, I found it very effective in getting KSEC to change their ways and improve their working practices. It has an almost immediate financial impact on a contractor and their sub-contractors and matters tend to get sorted out quickly. After once invoking the suspension clause, I found it only necessary to threaten further suspensions to get any other safety issues attended to without delay. This is just one of the many things that have been lost by dropping the role of Engineer in new forms of contract.

There was one rather serious safety issue where I threatened a further suspension. This involved a deep excavation for the new leachate sewer which ran beside the culverted stream at the bottom of the embankment of Tsuen Kwan O Road. In this location the sewer was around 8m below ground level in decomposed sandy granite. KSEC had sub-contracted the sewer work to a local contractor who had excavated a deep trench for laying the steel leachate pipe with a minimum of shoring on the sides of the excavation. George Anderson called me with his concerns. Again, like the KCR embankment issue, the problem had come to a head at a weekend, and I drove over to the site. I descended into the trench with George and saw there was a total absence of shoring on the bottom 2m of the trench. The sub-contractor explained that they had run into partially decomposed boulders and found it difficult to line and shore the trench as the boulders were in the way. I told everyone to get out of the trench and George and I went over to Mr Lee's office. Fortunately, he was in, and we took him back to the trench with us. I was reluctant to go down into the trench again, but I did so with Mr Lee and George. We stood at the base with sand trickling down, expecting the ground to engulf us at any moment. Looking up, I could see KMB buses on the Tsuen Kwan O Road, seemingly a long way above us. Mr Lee got the point and with some relief we climbed back up. I told Mr Lee and his sub-contractor that no more work was to be carried out at the base of the trench. The bottom two metres was to be backfilled and more robust shoring was to be installed in the upper portion. Only when this was done could the base be re-excavated to the required depth with more shoring all the way down, despite the presence of the boulders, which would have to be broken out individually as necessary. The alternative to this would be a further suspension. Things got sorted out fast and when I went back with George a week or so later to see the new pipe being installed, I was pleased to see the trench was firmly lined and braced from top to bottom in a workmanlike manner.

KSEC completed the contract with a delay of around one month. I awarded them an extension of time of the same amount with the agreement of HD on grounds of unusually inclement weather (a common ruse by a sympathetic Engineer to avoid imposition of LD) and in the absence of any major claims there was no significant cost overrun. I had got on well with Steve Riley and Peter Mak (and am still in touch with them both, even walking on Lantau with Peter). There was a feeling that we had made a good showing with our first commission. This was confirmed, as during the 12-month maintenance period for the works while we still had a presence on site, we were given the job of supervising installation of deep bored piles for the three trident blocks. We were confident then that we could expect further site formation work from HKHA.

We did obtain several other commissions, but only after some help from Dennis McNichol in getting us appointed as a registered consultant for the Authority. The requirement for a register was a new one being enacted by all government departments at that time. Things were becoming more bureaucratic and the old way of awarding work directly to consultants was now frowned upon. Unfortunately, in order to obtain registration with HKHA for site formation consultancy, an organisation had to have a minimum head count in Hong Kong and with our site establishment on the Trunk Road running down we fell short of the necessary number. One morning Dennis phoned up and explained this to me. However, he said he was keen to help us as we had done a good job on Kai Liu. We would need to form a joint venture with another firm in the same boat as ourselves—and he suggested Charles Haswell & Partners—so that the combined numbers would satisfy the new requirement. We could take the lead, he said, as CHP would deal with specialist geotechnical input only. Dennis said I should phone up Vic Turner who was CHP's partner in Hong Kong to agree the arrangements and that both of us should submit letters to him, that day if possible, confirming the firms would work together. I spoke to Vic who had by this time had got the same message from Dennis.

Letters advising of the formation of our joint venture were duly produced and delivered to the HD office in Ho Man Tin that afternoon. A day or two later I got a letter back from Dennis saying he was pleased to inform us we had been accepted on the HKHA register of consultants as the Fraenkel/Haswell joint venture. This forced marriage was successful and further HKHA site formation jobs followed as well as an unusual and very lucrative assignment a year or two later.

Travels in South China

After the death of Mao Tse-tung in 1976, the so-called "Gang of Four", which included Mao's widow, Jiang Qing, briefly took the reins of power in China, but were quickly deposed. Deng Xia-ping had been purged by Mao during the Cultural Revolution, but he came back to prominence and became the de facto leader in 1978. Deng soon instituted profound changes in China, combining China's brand of socialist ideology with the free enterprise policies of the West. Under him China started to open to outside investment and local companies in Hong Kong were quick to latch on to the opportunities which were arising over the border.

We got involved in some of these with a thriving shipyard business in Kwun Tong whose owner had an interest in investing in developments in Guangzhou Province. Possibilities included new container facilities on the Pearl River Delta, upgrading old shipyards on the river itself and providing temporary steel landing stages to be constructed in the shipyards in Hong Kong for use on coastal construction sites in South China. As Tony Bowley was working full time on finalising Trunk Road issues, Ted Foster asked me if I would look after this potential market in China and I gladly agreed. The opportunities were mainly in the marine sector in any case in which I had more experience from projects I had worked on in the UK. It was easy obtain visas for entry to China at that time. Hong Kong Chinese had a right of entry with their Hong Kong passports, and I simply used tourist visas with no questions asked, even though I was in China for business purposes.

My first trip took in Dongguan and Guangzhou in connection with a shipyard on the Pearl River near Dongguan. Our client was looking to form a joint venture with the Chinese shipyard to build construction barges and wanted a preliminary appraisal of the condition of the Dongguan yard and whether it could be upgraded at a reasonable cost. I was to be accompanied by the owner's son with whom I had had several meetings in our office and in Kwun Tong. After our meetings in Kwun Ting, he would drive me to the new MTR station in his father's Rolls Royce. I was always struck by the total lack of envy in the passengers as I got out of the Rolls and joined them in the queue at the ticket office. If you had made sufficient money in Hong Kong to buy a Rolls Royce, good luck to you was the attitude.

To get to Dongguan we took an early train from the railway terminus at Hung Hom and got off at Lo Wu where we crossed the border on foot. I had to go through a special custom's shed, but there was no delay and we proceeded into China. We were met by representatives of the Dongguan shipyard in an old long-wheelbase Land Rover and set off for Dongguan by road. In 1984 the roads in China were in poor condition and the single carriageway road to Dongguan and Guangzhou was no exception. Traffic was heavy in both directions, there were frequent delays, and it was nearly one o' clock when we got to Dongguan where we were to have lunch. It was the 15th of July and blazing hot. Dongguan styles itself the "lychee capital of the world" and the 15th of July is "lychee day," the very height of the lychee season. Not surprisingly we got a lot of lychees to round off the meal. What struck me was the great influence of Hong Kong. Old green light buses originally from Hong Kong were everywhere and we paid for our meal in Hong Kong dollars which the restaurant owner was happy to take instead of Chinese yuan. I was told this was illegal, but nobody cared. A strong currency was what was wanted.

In the afternoon we went on for a first look at the shipyard. It seemed quite dilapidated, but there were some vessels on the stocks and barges on its several slipways, although there was little activity in the hot afternoon. In the office we met the owner with the shipyard manager and had tea and local beer. My client discussed the possible business deal in Cantonese while I sipped the beer and looked out of the window at the yard. We stayed the night at a nondescript hotel in nearby Guangzhou, which struck me as a rather drab, grey place at the time. The next day I was back at the yard, with a view to assessing it properly, at least as far as civil engineering work for upgrading was concerned. We finished around 5pm in the afternoon, but rather than face another long road journey decided to fly back and managed to get on to a China Airlines flight that evening. In the office the next day I wrote up a report giving my assessment of what would be needed to bring the facilities to a reasonable condition. To augment my report, I recommended a visit by a mechanical/electrical engineer to assess the equipment such as lifting gear, winches and the electrical equipment. Now, over 30 years later, similar shipyards in and around Dongguan are producing sophisticated semi-submersible barges and other vessels of the highest quality while the yards in Hong Kong have mostly disappeared.

Next up for me was a trip to Zhuhai on the west side of the Pearl River estuary close to Macao. Now you would drive over the Zhuhai-Macao-Hong Kong Bridge which opened to traffic in 2018, but then we could only get there by ferry. This was to assess a site for development as a container terminal. In addition to the Zhuhai site our client wanted me to see an alternative site some way up the river. For this, he had somehow arranged for me to be picked up in Zhuhai by a launch crewed by members of the PLA. How this was arranged I had no idea although I assumed money had changed hands in fixing it up. The vessel was under the command of an army captain. He spoke very good English and pointed out various features of the river as we went. Still in sight of Hong Kong, we passed a distinctive island on which there were clusters of derelict wooden buildings.

"That is Lingding Island and what do you think these are?", the captain said, pointing to the buildings. Without waiting for a reply, he went on, "These are the godowns where your ancestors stored the opium", taking it for granted I knew of the opium smuggling and the war between Britain and China in the years before Britain took possession of Hong Kong. As it happened, I knew the basic details of the opium war and the role the Scottish adventurers and the local Mandarins played in the smuggling of opium which preceded it. I said I had read about that time, but I made no other comment except to say the island might be a good location for a container terminal, suddenly mindful I was in Communist China on a PLA cutter. Further upstream we came to the narrow strait of the Boca Tigris where the Pearl River discharges into the South China Sea and just beyond the strait, we cruised around other islands while I made notes and took photographs of possible container port locations. Later, I was dropped off back at Zhuhai and the captain and I parted on the best of terms exchanging contact details. It had been an interesting day.

My third foray was to Shekou which is close to Hong Kong. At that time, it was a town on its own, but now it has been absorbed in Shenzhen. We had lunch in a restaurant facing back across Deep Bay (*Shenzhen Wan*) to Hong Kong where I could see tall buildings in Yuen Long behind the Mai Po Marshes. For the first and only time I ate dog. The meat was quite bland, served in a spicy sauce and reasonably tender. I was more concerned when a dish full of oysters cooked in Chinese rice wine and soy sauce appeared. I was told these bivalves were very special and came from the best oyster beds in Deep Bay. At the time there were dire warnings in Hong Kong on the dangers of eating oysters from Deep Bay as they were said to contain excessive concentrations of heavy metals washed down the river from mainland China. I fished in the dish carefully with my chopsticks and withdrew just two small oysters which I declared delicious, but then turned my attention back to the dog, while everyone else wolfed the oysters down with relish.

After lunch we drove through Shekou town and stopped on the waterfront at a quay wall formed from steel sheet piles. There was no ground information to be had, but I assumed the piles had been driven through marine clay and probably some way into alluvium below the clay. However, they had clearly not been driven far enough as I could see clear signs of rotational slip failure of the wall and the filling behind it—a crack in the ground surface about 20m back with the wall leaning into the filling. Our client had an idea to use the quay for loading oil drilling equipment for use in oilfields which had been discovered near Hainan to the south. A major repair would be needed, probably in the form of a much more robust wall in front of the existing one, this time driven into weathered rock likely to lie below the seabed soils, and with a piled relieving platform behind the wall to carry live loading directly to rock. I wrote this up the next day in a report with a preliminary estimate of the cost based on Hong Kong construction prices. Not surprisingly, given the high cost of remedial works I came up with, nothing more was heard of the scheme.

My final trip in South China was to Daya Bay, just north of Hong Kong, where one of China's first nuclear power stations was to be built. Word of this had caused consternation in Hong Kong where China's ability to safely operate such facilities was called into question. This was before the Chernobyl disaster in the Ukraine, after which fear of the development became even greater. In fact, the plant was eventually constructed and was commissioned in 1994 with Hong Kong's China Light & Power taking 30% ownership. CLP's promotional blurb in 2019 says the operation of the plant has been specially commended within the nuclear industry for its exemplary safety record. So much for the fears in the 1980s.

Daya Bay lies up the coast of China from Hong Kong, beyond Mirs Bay and the New Territories. The idea was we would build steel pontoon units in the Kwun Tong yard which would be towed to Daya Bay to act as temporary breakwaters and construction jetties at the power station site. As I could see from the chart, the bay would be exposed to the full force of typhoon winds and waves and the design and mooring of these units would be difficult to say the least.

We travelled from Hong Kong in a Toyota Land Cruiser which had dual registration for Hong Kong and Shenzhen, crossing the border at Lok Ma Chau. Once on the China side, we drove parallel to the border on a narrow road for several miles looking into Hong Kong through the double wire fence that had been erected to deter illegal immigrants from mainland China, seeing Hong Kong as they would do before attempting their illegal entry. We moved inland, skirting around Mirs Bay and then headed back towards the coast again to reach Daya Bay. The last stretch of the road was very rough. Squads of labourers were knapping stones by hand to pave the road in front of us as had been done for hundreds of years. There were no excavators, spreaders or other mechanised plant and no asphalt surfacing, just hand tamping of the stone to form the road pavement. Soon we reached the sea at Daya bay.

The place was totally unspoiled with a beautiful golden sand beach sweeping in a gentle crescent to the north-east up the coast of China. I was used to the lovely beaches of Hong Kong, some relatively unspoiled, as at Tai Long Wan in the Sai Kung Country Park near where we lived, but I had rarely seen anything as pristine as the beach at Daya Bay. Also, the beach here was completely deserted unlike the beaches in Hong Kong which always attracted parties of young people, even in winter. It was difficult to imagine a nuclear power station in this idyllic place. We had arrived at the south-west end of the beach where there was a promontory separating Daya Bay from Mirs Bay and the New Territories in Hong Kong. This neck of land would offer some protection from typhoon waves and would provide a starting point for the breakwater. A lot of investigation would be required before any temporary breakwater/jetty scheme could be developed, but it would certainly be feasible to bring construction supplies and plant by sea if the contractors for the power station wanted to do so. Back in the office again I produced an outline design for discussion based on depth contours from the chart and such topographic detail as was available.

After this we reviewed our position on the potential for work in China. When we started working with the shipyard in Kwun Tong we had agreed to meet our own costs in the first instance in the hope of getting involved in major projects. The forays into China and the time I spent on the visits and in writing up reports had cost around HK$500,000. I now see it was naïve to expect to obtain work quickly and the best we could have hoped for was longer term associations which might have paid off in the future. It was great fun swanning around China, but we could not continue to fund promotion there indefinitely and reluctantly decided to let the association fizzle out. There was enough to be going on with in Hong Kong in any event.

Typhoon Ellen

Every year between May and October several typhoons threaten Hong Kong. They form initially as tropical storms deriving energy from the warm summer waters of Pacific Ocean east of the Philippines and usually track westwards towards the South China coast, sometimes intensifying into hurricane force winds as they do so. Many of these typhoons pass too far from the colony to cause problems, but from time to time, if conditions are right, Hong Kong can be hit by a severe typhoon causing extensive damage and loss of life. Earlier in the memoir I described the investigations we carried out to determine the optimum level of the Trunk Road to safeguard the carriageways from typhoon induced surges in Tolo Harbour. We were to experience one of the worst typhoons of the twentieth century and the effects of its surge during our stay in Hong Kong.

This was Typhoon Ellen which hit in early September 1983. Damaging typhoons in Hong Kong almost invariably have three main characteristics: they have very low barometric pressure at their centre or eye with high wind speeds in their circulation; on their way to Hong Kong, they track through the Bashi Channel only brushing Luzon in the Philippines and thus maintaining most of their energy from minimal contact with land; and their final track before the South China coast takes them just to the south of Hong Kong with winds blowing over open sea. Ellen ticked all these boxes and its passage also coincided with high tide adding to typhoon induced surge effects on sea levels.

Although it was hot and sunny in early September that year, as it often is when a typhoon is approaching, from the weather reports we knew a big typhoon was coming and on the morning of Thursday 8 September, the Typhoon Signal No.3 was hoisted by the Royal Observatory. This was followed by the Typhoon Signal No. 8 in the late afternoon. My wife came into Kowloon with all three boys in the car to pick me up. Traffic out of Kowloon was heavy with people leaving offices early and winds already gusting to gale force, but at that stage there was no rain. We taped up the large windows in the house to prevent injury from flying glass if they were blown in, much as people in Britain did against blast pressures during the blitz in WW2. Through the evening and overnight the wind continued to rise from an easterly direction. At some time during the night the No.10 Signal was raised, then the No.11. By Friday morning the storm was at its peak. The air was full of branches torn off the trees and other debris sucked from the ground in periodic wind gusts of great intensity. These appeared to come around one minute apart. I had parked the car in the garden area beside the house as far away from trees as possible and looking out at the peak of one of the gusts I saw the car move forward several feet as the gust caught it. The same thing happened with the next gust, and I realised the car would soon be blown to the end of the garden where it would go over a slope to a path below. I would have to move the car to the front of the house out of the worst of the gusts, although that would bring it closer to trees. Waiting until a gust had just passed, I rushed out, started the car and quickly moved it round. As I got out of the car, I could hear the next gust coming above the general roar of the storm and ran back into the house and closed the door just in time. For that minute outside, it was like being in a different world where the norms of nature as I knew them did not apply. I felt an atavistic need to get into shelter as quickly as possible. Anyone out there for any length of time would have been lucky to survive amongst the whirling debris of the storm. Indeed, in Hong Kong that day eight people were killed and many more were injured.

Around lunchtime the wind started to ease and by 2pm it became almost calm. I went out and saw blue sky above—we were in the eye. Soon the wind picked up again, now blowing from the west, but with less strength than before. During the evening, the Royal Observatory reduced the typhoon signals to No.8 and then to No.3 and the storm dissipated on reaching land south of Macao.

The road to Sai Kung was blocked with fallen trees and we were not able to get out of Long Keng on Saturday. By the Sunday, the trees had been cleared sufficiently to get to Sai Kung, driving with some care. There were numerous unofficial moorings In Port Shelter north of Sai Kung. We could see that some of the vessels at these had been driven off their moorings by the surge and swept right across the road some way inland and wrecked in the process. As the moorings were unofficial, and the vessels were not in a designated typhoon shelter, there was no insurance cover for them. It was estimated the surge height was around 3m in Port Shelter on top of the high tide. Hurricane force winds had been recorded in all parts of Hong Kong, with the south of Hong Kong Island worst hit, experiencing wind gusts around 250km/h. We were less exposed than most as we were surrounded by wooded hills and suffered no damage. it was an interesting experience, nevertheless.

The Joint Declaration

Other turbulent events were brewing in the background, although these were political, rather than meteorological. When I first went to Hong Kong in 1977, there was some speculation on what might happen at the end of the New Territories lease in 1997, twenty years hence. There was a suggestion that any uncertainty might lead to a loss of confidence in the financial markets resulting in a flight of capital from Hong Kong and pressure on the dollar. These thoughts were quickly forgotten, however, and people soon got on with their lives in the Hong Kong fashion. But in 1982 the subject came up again when it was announced that the UK prime minister, Mrs Thatcher, would be visiting Beijing in September and would meet with Mr Deng. Why she went there at that time is not clear, but undoubtedly the future of Hong Kong was discussed at the meeting, although little was said at the time. Perhaps, buoyed by the British victory in the Falklands, she thought she could extract a favourable future arrangement for Hong Kong out of Mr Deng. The expectation of most of the British and Chinese in Hong Kong was that the lease would be renewed, perhaps for another 50 years, or failing that, there would be some form of condominium with joint administration by Britain and China. This proved to be a vain hope. After Margaret Thatcher's visit, meetings were held between a British delegation and Chinese government officials, with involvement of the Hong Kong Governor Sir Edward Youde. From these it became clear that our expectations were just wishful thinking and far off the mark. The Chinese government position which was emerging from the talks was that any continuation of British administration after 1997 would not be possible and the whole of Hong Kong, not just the NT, must be returned to China. Not surprisingly, in the autumn of 1983 when this leaked out there was consternation with a run on the Hong Kong dollar. In the space of a week the dollar's value fell from around 8 to the pound to over 16 and continued to fall.

Fortunately, there was a cool head in the colonial government, the financial secretary, John Brembridge. He was not a career civil servant, but had held a senior position in John Swire, before taking up the government post (incidentally, our secretary Daisy Chan had worked for him in Swire). It was Brembridge who took the controversial step of pegging the Hong Kong dollar to the United States dollar at the rate of HK$7.8 to US dollar. This stabilised the currency, but at the expense of driving up interest rates to unprecedented levels of over 15%. It did not take long, however, as people came to terms with things, for interest rates to fall back to normal and Brembridge's sangfroid undoubtedly saved Hong Kong from financial disaster. In fact, the peg endures unchanged to this day.

All this also affected our own finances. My wife and I had a joint mortgage on our house in the UK and had to transfer a slice of my salary every month to pay the instalments. I was paid in Hong Kong dollars and immediately after the peg was introduced it cost us an extra 50% in dollars to service the mortgage. Fortunately for us, the pound soon fell in value against the US dollar and after a few months we were back to transferring roughly the same amount as before.

Negotiations between Britain and China continued in order to agree the final terms of the handover when the lease ran out on 1 July 1997. While there was no longer any prospect of continued British involvement, the two governments reached an agreement that after 1997 Hong Kong should have a high degree of autonomy from China and its rights and freedoms should remain unchanged for a period of 50 years from the handover. This became known as the Sino British Joint Declaration which was agreed in 1984 and registered as a legally binding treaty at the UN in June 1985. It was not, perhaps, what the majority in Hong Kong would have wanted, but it was still a triumph for the negotiators and with the stability it offered, Hong Kong continued to prosper up to 1997.

The Kowloon Cricket Club

I had applied to join the KCC shortly after arrival in Hong Kong in 1982 and was surprised and pleased to be offered membership within two years. This made life so much better in many respects, particularly as the club was very close to our office in Tsim Sha Tsui. The KCC had been formed in 1905 when it acquired a lease of British War Department land for use as a cricket pitch and the first pavilion was built in 1908. The present pavilion was opened in 1932, although it had been extensively modified when I joined in 1984. As well as cricket, the club offered hockey, bowls, tennis and squash, several restaurants, a bar and a swimming pool. But the one feature above all which distinguished the KCC was its large cricket field set in the centre of Kowloon, lined with trees and bushes—a green lung in one of the most highly populated cities on earth.

We had family membership and the older boys had tennis lessons and played squash with their friends. They could travel to the KCC after school taking the MTR to Jordan station close by. Margo would often drive to the club with Stephen, and I would walk over and join them all after work, either to eat there, or to drive home with them to eat in Long Keng. Margo and I also made good use of the upstairs grill in the club which served excellent food. Almost every day I would play squash at lunchtime, mostly with Tony Bowley (who enjoyed the luxury of membership, not only of the KCC, but of the USRC as well) and sometimes with George Anderson who would come over from the site at Kai Liu. Around once a week, I would play squash in the evening for one of the lesser KCC teams against teams from other clubs and I also played for a team formed of KCC members called Tsing Tao which was sponsored by the Chinese brewery company, against mostly Chinese teams from around Hong Kong. These matches took us all over the colony and we usually found somewhere to get beer afterwards and food from a *dai pai dong*.

One evening I was due to play for a KCC team against a USRC team against whom there was always some rivalry as the two clubs were located adjacent to each other. I had to go out to a site all day and suddenly realised I had forgotten to pack an essential item in my kit. I knew my wife would be coming into Kowloon that afternoon and phoned her to ask if she could get a replacement item for me in a sports shop and leave it at the reception desk in the Club to collect when I came for the match. She was not keen, but eventually agreed and I described what to ask for in basic *Gweilo* Cantonese. I called again when she was back home and was assured the newly purchased item would be waiting for me at the desk. She said, however, that when she had asked for *Yat goh jock strap* she was puzzled at first when the assistant placed not one, but three on the counter. But, as the assistant explained to her in perfect English, jock straps came in three sizes—small, medium and large, which I had forgotten to say. When I got to the changing room, I was pleased to see she had picked the correct size.

House and Garden

We were happy with our rented accommodation in Long Keng where we had the use of a sizeable garden area beside the house in which the boys could play football and other games such as badminton (on calm days). The garden did not belong to our landlord as it was still government land. The Small House Policy only entitled the long-term NT resident sufficient land to build the house on the 700 square foot plot plus access paths not exceeding 5 feet all round. Nevertheless, we treated the area like a garden and kept it tidy and cut the grass. Then Mr Wong announced he was going to build a house in that area. No doubt, some other long-term NT resident would be named on the title deeds as far as the land registry was concerned, but Mr Wong would be the real owner. I objected and said one of the main things we liked about the house was its open aspect with the garden and we might have to move. While the Chinese generally have less interest in gardens than the British, he no doubt appreciated the attraction of the property would be reduced by having another property next to it cutting off the light, and perhaps also diminishing the *fung shui* and he offered a solution. We could move to the new house when it was completed—same rent, no problem! He would even create a bigger and better garden next to it. I agreed and in a remarkably short time the new house was built. It was superior to the one we were in with more expensive fittings and seemed bigger all around with bathrooms on all floors. There was a veranda which looked out across the old paddies as with our existing house, but much bigger. My wife even persuaded Mr Wong to provide a large retractable canopy over the veranda which we could run out during heavy rain or as protection from the summer sun. After clearing some small trees and bushes, Mr Wong constructed the new garden area with the aid of large quantities of imported filling and built a wall to retain it in line with the veranda where the ground fell away steeply.

This retaining wall caused him some problems, however. One evening when I had just got home from work, he came to me and asked for my advice as an engineer regarding the wall. It was over 4m high and while the wall stem contained steel reinforcement, it was no more than 150mm thick and its foundations were inadequate. As a result, it had suffered some cracking and had tilted forward and the whole thing looked to be on the point of collapse. I told him he had two choices: either knock it down and build a properly engineered wall or try to buttress it. Buttresses would be the quick solution, but they would need good foundations and enough mass to resist the sliding forces from lateral soil pressures without any help from the wall itself. I did some rough calculations and sketched out what he needed to do, emphasising the importance of the foundations for the buttresses and the need to cast them firmly against the face of the wall. I also warned him to keep the ready-mix truck at least 5m back from the wall during the concreting operations. I kept an eye on things as the foundation excavations were completed and rough shuttering was erected. Mr Wong and his men worked flat out and soon they got concrete into the forms. After a few days, which was probably longer than necessary in the warm weather, I said to strip off the shuttering and the three buttresses were revealed looking quite workmanlike. There was no more movement—everything was stabilised. In 2015 I had occasion to walk along the old paddies with my friend Mark Cheung, to whom I pointed out the wall and its buttresses, and I am pleased to say the wall was still standing after 30 years.

As we were getting the keys to the new house from Mr Wong and dealing with the paperwork for the rental agreement (at the same rent as promised), I complimented him on a job well done. Then, out of the blue, he offered to sell the new house to me. The price would be HK$500,000. I could have got a mortgage quite easily from the Hong Kong bank with monthly payments no more than my rent allowance, but, stupidly, I did not take up the offer. This was a big mistake. When we left Hong Kong three years later Mr Wong sold the house to a Cathay Pacific pilot for HK$2 million and in 2012 when my wife and I were in Hong Kong and took a trip out to Long Keng it was on the market for HK$24 million. But even more than the lost investment opportunity, I doubt if we would have left Hong Kong at all if we had owned the house and the difficulties to come for us within PFP would have been avoided. Of course, that was all in the future, and I had no way of knowing the way things would work out.

Beyond the Trunk Road

With completion of the Trunk Road, my focus was mainly on the HKHA projects which we were now tackling, with CHP dealing with the geotechnical issues. Joint ventures and associations between consultants often break down, especially when they are formed with the objective of securing new work which may not materialise, but this joint venture with CHP for HKHA projects worked out well. Charles Haswell in some ways had a background which was not unlike that of Peter Fraenkel. Like Fraenkel, Haswell had been a partner of a large London consultant, in his case, Halcrow, and had left to start his own firm. Charles Haswell's specialty was tunnelling, and this found a ready market in Hong Kong with new MTR lines and water tunnels. In addition to the HKHA work, we did some other small jobs with CHP. One of these was a preliminary study in Singapore on a section of the Mass Rapid Transit metro planned in Singapore following the success of the system in Hong Kong. Haswell had taken on an ex-Director of the Hong Kong PWD called Gordon Sapstead who was leading the CHP team on the Singapore job. It was from him I learned of the idea the PWD had in 1977 of PFP becoming the third force for NT highways schemes to reduce reliance on Maunsell and Scott Wilson, which opportunity was let slip.

Despite the arrangement with CHP, it was to us alone that Dennis McNichol came to with an unusual assignment regarding repairs to high-rise housing blocks, presumably because it involved structural issues which were not CHP's field. I was aware there had been a recent scandal in construction of some of the housing blocks in the new estates being built all over Hong Kong. It seemed that employees of certain contractors, with the connivance of some HD supervisors, had deliberately reduced the cement content of the concrete mixes the HD had specified for the housing structures, while still being paid at the full rates for the work. In some cases, the strength of the concrete in the structures was as little as 20% of what had been specified. When this was discovered, there were criminal proceedings and some convictions, but the damage had been done and a significant number of housing blocks were dangerously understrength and could not be used, putting serious pressure on the supply of public housing. Some of these damaged blocks were demolished, but HD considered that others could be saved with strengthening measures and a programme of repairs organised by their own structural engineers was put into effect as a matter of urgency. Dennis asked me if we could take on responsibility for supervision of some of this work, recruiting site engineers and inspectors for site duties. He made it clear he had not asked any other consultant for help and if we could accept the charge rates for the staff which he could pay we could have the work straight off with no competition. Almost, apologetically, he proposed a mark-up of 25%. Used to the 7.5% mark-up on site staff costs paid by the NTDD, I accepted on the spot.

Someone recommended a retired Chinese engineer who had been RE on housing construction in the past who would be interested in taking a role in this work. I interviewed him, liked him, and took him on right away to manage the staff we would recruit. I brought in a few of our graduates to work under him as section engineers along with some of our ex-inspectors of works from the Trunk Road. With this cadre to start with the team soon expanded to around 20 in total. I had little to do in management other than countersign the monthly reports and submit invoices to HD. With minimum management effort we were soon making a regular income of close to HK$800,000 a year from the appointment.

Around this time Ted Foster announced he was retiring and returning to Australia. He had managed the office for about 5 years in a pro-active manner missing under his predecessor. One thing Ted did before he left was to pass on his company car to me for which I was most grateful, as the car I had bought on arrival in Hong Kong now needed a lot of attention and was always in the garage for repairs. Ted's car was a Toyota Celica in excellent condition. As my terms did not run to a company car, I paid a nominal sum for it, and it served us well for the remainder of our time in Hong Kong. While the Trunk Road had been largely completed and the regular income from it had ceased, a wide range of smaller jobs had been obtained so that the office was running profitably. Tony Bowley and I were made full partners, albeit with a miniscule share of the profits and a Chinese engineer called Peter Chan was brought in to try to expand work with local private sector clients. Tony and I worked well together. Highways and traffic engineering were in Tony's province; I dealt with bridges and other structures as well as geotechnical issues, although for these I was happy to bring in bring in specialists such as Len Threadgold or CHP's geotechnical engineers when necessary. We had become very good friends and Tony's wife Sue, was friendly with Margo. Although Tony was more outgoing than me, we moved in roughly the same social circle centred on the KCC's sporting activities, and of course, the bar, and we had many good times together.

Stock Market Crash

By the mid to late 1980s Hong Kong people had come to terms with the fact that the colonial days were coming to an end and for better or worse China would be running the show after 1997. Once financial fears had subsided and the dollar peg was seen to be working, the economy boomed. More and more banks and other financial institutions were setting up in Central, property prices rose rapidly, and the stock market surged. Hong Kong took advantage of its status as a free economy on the edge of the massive Communist state over the border which was now starting to shake itself free of restrictions under the liberalising influence of Mr Deng. Manufacturing businesses of all kinds in Hong Kong were shifting their production facilities into China where wage costs were a fraction of those in Hong Kong with fortunes being made on both sides of the border. Hong Kong was also seen as a haven for newly acquired Chinese money as it had ready access to international financial markets and other hard currencies. "Templar" who produced a daily strip cartoon on topical issues for the *South China Morning Post* put his finger on the way things were, with a few modifications to Kipling's famous verse:

Oh, East is East, and West is West

And ever the twain shall meet,

As they push and shove on the MTR
Or jostle on Pedder Street
But there is neither East, nor West,
Border nor breed nor birth,
When everyone is milking the place
For every dollar its worth.

There was a rude awakening in the autumn of 1987. Out of the blue in October of that year stock markets around the world were hit by a sudden crash in share values. It is thought that the falls were triggered by computer program trading models which had recently been introduced, followed by panic in the investors. Both London and New York fell by around 25%, but in Hong Kong where shares had become grossly overvalued, the fall was 46%, the most of any major market. Trust in the Hong Kong stock market also took a knock when Ronald Li, the then Chairman of the stock exchange, suspended trading on the plunging market "to protect the investor". It took years for shares in Hong Kong to regain their value, but as usual in these situations, those who bought at the low point were rewarded handsomely in the end.

The stock market crash made little difference to us as we held no shares at that time and while there was damage to pension funds, the need to draw on my pension was still a long way off with plenty time for recovery. There may have been some slowdown of the construction industry, but Hong Kong was nothing if not resilient and as PFP's workload was still mostly composed of long-term government projects we were not affected.

Activities Outside of Hong Kong
Sabah, Sarawak, Thailand and Bhutan

For several years in the 1980s, I rarely left Hong Kong, except for the promotional forays into South China, holidays with the family in Penang and occasional home leave in the UK. But after some time focusing entirely on work in Hong Kong, there were calls for advice from me relating to the bridges in Sabah and Bangkok which I had designed, and which were now under construction. Some things could be dealt with without leaving Hong Kong, but other issues needed visits to the sites. I also undertook promotion with trips around S E Asia and to the Indian sub-continent.

Sabah

In Sabah, only two of the routes we had designed had gone ahead to construction with World Bank funding, Sandakan to Telupid and Sandakan to Lahad Datu, as the western routes had been deemed not economically viable. Incredibly, while funding the road between Sandakan and Lahad Datu, the World Bank had also refused to finance the large bridges saying the cost/benefit ratio of replacing the ferries did not meet their economic criterion. This was madness as what was the point in spending millions of dollars upgrading the road if travellers between the towns could be stuck for days at the unreliable ferry crossings. However, the Malaysian government, seeing the pressing need for the bridges, had struck a countertrade deal with the Indian government involving construction of the bridges by Indian contractors in return for supplies of palm oil and other commodities. In keeping with these financing arrangements, Minco were now the lead consultant for the construction phase of the project, and we were reduced to a subsidiary role.

Minco sought advice from me on the piling for the foundations of the main piers of Kinabatangan Bridge and I was pleased to travel to Sabah to help. The site investigation at the design stage was now being augmented by further investigations commissioned by the Indian contractor, mainly to determine the length of piles which would be required. We had known from what had been done previously that the ground was poor in engineering terms, being a river flood plain. I had opted for 1.0m diameter steel tubular piles which I expected could be driven through typical flood plain deposits to end bearing on limestone rocks even if these were at some depth. However, the site investigation was showing the deposits to be much deeper than expected with the rock over 60m below ground level in some places. Minco were asking whether the piles could be driven that far and if not, should the foundation design be changed. To answer this, I needed to see more details of the soil layers above the rock. I went to KL to discuss the S.I. results with Minco staff and then flew on to Sandakan from where I was driven to the site by Minco's RE to see the latest driller's logs. We arrived at the site to find the rigs standing idle and nobody about, with the ferry and its operators on the other side of the river. Minco's RE said the drilling crew were staying at a kampong in a place called *Bukit Garam* (Malay: hill of salt) about 5 miles up-river, so we hired a boat with a boatman and went to find them. When we got to the kampong, we spoke to the headman and asked if he knew where the drilling crew were. His response was not what I had expected. "All in hospital, Tuan", he said, "Malaria".

At this I felt something on my arm and quickly slapped a mosquito which left a smear of blood on my skin, too much to be just from me. Thanking him in my broken Malay we returned to the site on the boat. I realised I had completely forgotten to take any chloroquine or other malaria prophylactic, having got out of the way of doing so in Malaria-free Hong Kong. Several large gin and F&N tonics were downed that evening in my hotel in Sandakan. I do not know how much quinine I got from these, but they seemed to do the trick as I did not go down with Malaria.

Eventually I received full details of the boreholes which showed alternating bands of sand and clay to a depth of about 40m below ground level under which there was a layer of hard clay above the rock. From the soil strength values indicated by the SI, I calculated that the piles would not get through the hard clay to reach rock even using a very large piling hammer. However, with the piles driven to about 50m down, they could carry the design loads as friction piles. Minco seemed happy with this advice and a variation order was issued for the revised design.

A bit later I made another trip to Sabah to investigate a problem which had arisen in Segama Bridge. This was potentially more serious. Soon after construction of the bridge transverse cracks had appeared on the underside of the concrete deck on either side of the north pier and water was dripping from them. Minco and JKR were concerned there might be a structural problem with the design. I went to KK and then flew on to Lahad Datu early the next day and spent some hours at the bridge (well dosed up with chloroquine this time). From underneath, I could see the cracks clearly, but noted they were only present at the north pier with none at the south. As the bridge was completely symmetrical about the centre of the main span this suggested the cracking had not arisen from a design fault. I gradually pieced together what had happened. In continuous composite steel and concrete bridges, the concrete in the deck above the piers is in tension. Concrete has little tensile capacity and anti-crack steel reinforcement is provided to counter the tension and limit the width of cracks which form. To further limit the crack width, concrete over the piers should be cast only after the concrete at the middle of the spans to reduce the tension stresses as far as possible. Despite having spelt this out clearly in the specification I had written, the contractor had simply started concreting at the north end of the bridge and worked his way across from end to end. This meant the hardened concrete at the north pier had been stressed more highly than elsewhere in the deck. It also explained why there was no cracking over the south pier. This had not reduced the ultimate strength of the bridge, but it had caused greater cracking than expected. However, I calculated the crack width to be 0.18mm which was just within the maximum crack width of 0.2mm allowed in the British code BS 5400 which JKR used.

Although I could not make the cracks go away, I realised I could make them less noticeable by eliminating most of the water leakage through them. I had specified no-fines concrete in the footpaths. This is a concrete composed of only cement and coarse aggregates which is easier to break out if service ducts need to be laid in the footpaths at a later stage. No-fines concrete had been in fashion in the UK and Hong Kong in the 1970s, but it had fallen out of favour since then. As it is porous it was allowing rainwater to percolate through the footpath and into the cracks making them more visible from below. Probably also in the long term this would lead to unnecessary corrosion in the steel reinforcement. It would be best, therefore, to replace the no-fines with ordinary concrete.

Back in KK I wrote up a report for Minco and went to the JKR office with Minco's representative to discuss my findings and recommendations. I met the Assistant Director whom I remembered as being one of the nippy little forwards in the JKR football team we played in 1980. This tempted me to mention the ringers from the State team, but I thought better of it as you can never be sure of a client's sense of humour. He accepted my recommendation to do nothing except replace the no-fines. The bridge is still happily doing its job over the river.

Sarawak

I first visited Sarawak in the company of the manager of the Malaysian office of John Howard, a British contractor with offices in the East (now out of business). This was to prepare tender designs for three bridges: one on a tributary of the Rajang River near Sarikei in the Second Division of Sarawak about 200 miles from Kuching, one over the Limbang River in the Fourth Division close to Brunei and one at Lawas in the Fifth Division in the far north of Sarawak.

The crossing at Limbang was the largest at about 120m and I proposed to use a slightly cut-down version of the Kinabatangan Bridge. Limbang had been occupied by rebels in the Brunei revolt in December 1962 which had spread out into Sarawak and led to the fighting between the newly formed Malaysia under Tunku Abdul Raman, aided by British troops, and Sukarno's Indonesia a little later. The revolt in Brunei itself was quickly put down by the Queen's Own Highlanders and the Gurkhas. In Limbang, where the hard-core rebels were concentrated, the fighting was more intense, but the town was retaken by two companies of 42 Royal Marine Commando. There is a well-tended monument in the *padang* to several British soldiers who lost their lives in the action.

After Limbang, we flew up to Lawas which had a short airstrip surrounded by forest where the pilot had to make a tight turn around a wooded hill on landing or take off. After inspecting the site proposed for a new bridge where the river width suited a steel composite structure with a span of 25m as used in Sabah, we had lunch in a *kedai* then waited at the airstrip for the afternoon plane which would take us back to Kuching. I was booked to go from there to Singapore and then on to Hong Kong while my companion from John Howard would go to KL. As we waited in a little terminal building it started to rain and this increased steadily into a full-blown tropical downpour accompanied by flashes of lightning with thunder reverberating around the hills on all sides. It did not look good for our flight. We could hear an aircraft above the storm, but then the sound faded away and the controller came out to tell us the pilot had taken one look and headed back to Kuching: a hotel of some sort would have to be found with the hope of getting out the next morning. Then we had an idea—why not try to get to KK in Sabah and fly out from there? We were close to the Sabah border after all and there was no guarantee we would get out from Lawas the next day as the weather might be problematical for some time. When I worked in Sabah there was no proper road to Sarawak, but I had heard an all-weather road was being built. We decided to give it a go and asked the terminal to call a taxi for us.

A taxi appeared and the driver confirmed he could get into Sabah, although he quoted a high price for the 150 miles to KK saying he would be unlikely to pick up a fare for his return journey. First, he would have to let his wife know what was planned and then he would take us. It was still raining when we eventually set off. After about an hour the road got very rough. There was a road formation of sorts, but it had not yet been surfaced and there was a lot of standing water on parts of it. Then we came to a barrier with a small building beside the road which our driver said was the customs post between Sarawak and Sabah. By now it was completely dark and still raining heavily. We waited, but as nobody came out, he edged the taxi past the barrier, and drove on into Sabah. Soon the road improved, and we came to Sindumin which I remembered from my work on the Sabah roads project and then on through Beaufort and finally to KK. It turned out it was a holiday weekend in Hong Kong which I had forgotten about, and KK was very busy with people who had come over for the weekend, but eventually we found rooms in the Capitol Hotel in the middle of town. The next morning, we fixed up flights to our respective destinations, in my case direct to Hong Kong. At such short notice I could only get a first-class seat on a Cathay Pacific flight, but I was not complaining. Our flights were called at around the same time and we made our way to security and presented our passports. This caused some consternation—we had no Sabah entry stamps and the eastern states had individual entry rules unlike the western states which had all been part of the old Federation. But how did you get here, we were asked by Passport Control. We explained and in the usual good-natured Sabah way, the Sarawak stamps were deemed good enough, and we were waved through. There was beer in the departure lounge, and we drank to the success of the bid with a couple of quick bottles of Tiger.

In the event John Howard's bid was not successful, but at least we received the agreed at-cost fee for our efforts and out of the visit another opportunity for bridge work in Sarawak came up. This involved a proposal for the design of six river bridges in different divisions of Sarawak. It was a government job with JKR in Kuching as the client. The project was for the replacement of existing narrow steel truss and timber bridges on roads which were in the process of being upgraded to all-weather standard. We would need local consultants to partner with. By this time Patrick Augustin and Inbaraj Abraham had left Minco and set up their own firm of consulting engineers, along with a Malay engineer called Faisal who provided the necessary bumiputra presence, but who was a good engineer in his own right having studied at Strathclyde University in Glasgow. They operated under the name of *Perunding Faisal, dan Abraham dan Augustin Sdn. Bhd. (PFAA)*. We joined up with PFAA for the bid, starting a long-term relationship which lasted over 20 years. Patrick arranged for a consulting engineering firm KTA who were based in Kuching, to come in as well. I flew to Kuching to meet Patrick and a partner of KTA, Walter Sim, to get a feel for the requirements, draft a proposal and to meet the local JKR.

I spent some time in Kuching discussing our approach to the work and preparing a first draft of the proposal. In the 1980s Kuching was a pleasant place in which few buildings were over 4 storeys, other than the Holiday Inn where Patrick and I stayed. Unlike KK, Kuching did not appear to have been damaged in the war and it had a strong colonial feel to it. There are conflicting theories about how Kuching got its name, whether from the Malay for cat, *kucing*, or from a local fruit, *mata kucing*. For over 100 years it was run, and run very well, by James Brooke and his descendants. Brooke was a British adventurer to whom Sarawak was ceded after he defeated the Iban head-hunters who were plaguing the coastal settlements.

Patrick had to go back to PJ while I was still working on the proposal, but I got on well with Walter and we finished the draft together, based mainly on using the type of composite steel and concrete bridges which had been successful for us in Sabah. He was very hospitable and took me around the town as well as to restaurants on the Sarawak River. He also got me invited to an "open house" lunch at the home of the *Mentri Besar* who was keen to learn the latest developments in Hong Kong.

I had expressed an interest in seeing inside a longhouse before I left and Walter arranged a memorable trip about 50 miles up the Sarawak River to a little settlement on its banks where he was known, having installed a reliable water supply for the settlement a year or two before. We powered up-river at over 30 knots on a narrow launch with a large Caterpillar engine. At the settlement, we were met by the headman at a little landing stage and were ushered over to one of the three longhouses in the settlement and then up a notched log set at an angle of about 40 degrees which served as access to the building which was supported on long stilts high above the ground. In the longhouse we sat in a sort of reception area where we were served tea and some fruit and chatted in a mixture of Malay and English. I mentioned the old practice of the Ibans of head-hunting and shrivelling heads and wondered if they still had any shrunken skulls around from those days. The headman smiled at this and conducted me along a central corridor past dozens of separate family-living and cooking areas to the far end of the longhouse. He pointed to a line of black objects hanging from a rail which on close inspection were recognisable as human skulls, probably about half the original size. The skull at the far end of the row was sporting a pair of cheap, metal rimmed glasses. The headman pointed at it and cackled, *orang Japan*.

(There is a footnote in the book *Flashman's Lady* by George MacDonald Fraser, which is partly set in Sarawak, on the same subject. The book features Flashman assisting James Brooke in battles with the Ibans and when researching for the book, George MacDonald Fraser describes a similar incident in a longhouse up the Rajang River. In his case the row of heads looked comparatively recent. Although the Ibans said they were mostly from *orang Japan*, taken during the Second World War, he thought some looked new enough to have come from Indonesians fighting British-Malaysian forces in Borneo in the border conflict between Malaysia and Indonesia of the 1960s, the "Confrontation". Old habits die hard).

Then I found it was one thing to have skipped up the notched log to the longhouse, but quite another to go down. There were no handrails or other supports, and it required an effort of will taking each step down the steeply sloping, narrow log with the ground seeming a long way below me. Walter led the way and was down in half the time it took me, and I was followed by an old lady who came down in a matter of seconds. Once down I was shown how to propel a dart from a blowpipe, or *simputan*, which is quite easy and very accurate when you know how, and presented with a blowpipe as a souvenir, now hanging up on the wall beside me as I write.

Thailand

Construction of the Rama IX Bridge in Bangkok was well underway by 1986 and I made several visits from Hong Kong to inspect the approach viaducts which I had designed. I also took the opportunity to see some of the erection procedures for the cable-stayed bridge.

The approach structures on either side of the river were put out to tender as separate contracts. As it happened, both contracts were won by a consortium headed by Maeda Construction Co. Ltd, a Japanese contractor active in Hong Kong, for whom a friend of mine in the squash team, David Westwood, was working at the time. Through David we were asked to check the structural design of the erection gantries, at a fair fee of course, which suited us well in view of the preliminary design I had done for the tender drawing set.

Maeda constructed two identical gantries, one for use on each side of the river. With 13 spans on each side and separate structures for each 3-lane carriageway this gave 26 uses for each of these gantries in two runs of 13. The gantries were 108m long and weighed over 900 tonnes. They comprised three steel box girders each with "noses" and "tails" in the form of steel trusses for launching and tie-down purposes respectively, with heavy rails on the underside of the box girders which ran over rollers on the pier tops. After casting and stressing a span, the shutters were withdrawn by hydraulic rams and the whole gantry was lowered on to the rollers. The gantry was then moved up the approach structure slope by another set of rams reacting against the end of the newly cast deck and the process was repeated. This erection system was very successful, so much so that the contractor was soon achieving a cycle time for completing a 50m span of only two weeks instead of the three weeks which had been programmed for.

The trickiest part of the process was at the last deck of each carriageway where the main bridge superstructure was already erected at what we called the "Junction Pier" and obstructed the gantry noses. Maeda came up with a clever way of dealing with this. Each nose was made in several segments 8m in length which had bolted joints between them. Temporary support was provided about 9m from the Junction Pier, so that 8m segments of the nose could be safely unbolted one at a time as the gantry was advanced across the temporary support for casting the last span. A foundation for this support would have required costly piling to support it in the weak ground, but Maeda avoided this expense by devising an inclined tower on knuckle bearings seated on the Junction Pier pile cap. The tower was stayed by a steel tie to the top of the pier, and not surprisingly was christened the "Pisa" Tower.

There was still one difficulty to overcome, however, which was getting the main parts of the gantries down to the ground with access for cranes blocked by the concrete deck above. For this, Maeda engaged VSL of Switzerland who were specialists in heavy lifting equipment (VSL also provided all the prestressing components). In each case, the whole gantry was first suspended on steel cross beams on the newly cast deck with support cables passing outside the deck outstands on each side. The two side girders of the gantry were detached from the centre girder and lowered on to ground supports simultaneously, in order to preserve balanced loading on the cross beams above. Getting the centre girder down was more difficult, especially the parts at the piers and required more temporary steelwork fitted under the deck to support the cables. Where a gantry was to be reused for casting the adjacent carriageways it was important that these dismantling operations were accomplished as quickly as possible. By virtue of clever engineering, Maeda and VSL achieved this and both approach bridges were completed well ahead of programme.

The cable-stayed bridge contract was won by Hitachi Zosen Corporation leading a consortium, including amongst others Kobe Steel Ltd. Their bid was based on carrying out the steel fabrication in Thailand, which had been dismissed as impracticable by other bidders. In fact, this gave Hitachi Zosen a competitive edge as sufficiently skilled local fabrication resources were available at significantly lower cost than in Japan or other industrialised countries. Then, just as the main bridge contact was starting, Dr Homberg resigned in a dispute over fees with the client. This left the remaining firms in our consortium in a difficult position as we had lost his big bridge erection experience. To get over this we engaged Freeman Fox & Partners (FFP), who were still regarded as the foremost bridge engineers of the day, despite their problems at Milford Haven and Yarra, to check the safety of the erection methods proposed by the contractor and provide other advice as necessary relating to the bridge.

The main piers of the bridge, termed the "Pylon Piers" as they supported the cable pylons as well as the bridge deck, were located on the banks of the river and were constructed within sheet piled cofferdams. They were founded on 2m diameter concrete bored piles cast under bentonite (a special drilling mud of high specific gravity used in piling operations). The contractor opted to use a reverse circulation drilling method, but trial piles failed loading tests. It appeared this method was causing disturbance and softening of the dense sand layer at the base of the shaft on which the piles were founded, and the piling method was changed to an auger and bucket method combined with pressure grouting of the toe zone of the piles. Further load tests on piles were carried out successfully including one to a test load of 2,000 tonnes.

A new fabrication facility was set up from scratch downriver from the bridge which produced 15,500 tonnes of permanent steelwork for the deck and towers together with 1,500 tonnes of temporary steelwork. Suitably experienced welders, mostly recruited locally, were tested and accredited for the various classes of work used in fabricating the bridge steelwork including submerged arc and semi-automatic gas shielded welding. Completed deck segments for the bridge weighing up to 200 tonne each were transported from the fabrication yard to site by Chao Phraya River barges. Erection followed the general method envisaged by Dr Homberg: firstly, the anchor spans for the back stays, then the pylons and finally the main span working out as stayed cantilevers from either side of the river. High strength friction grip bolted joints were used between the segments which came in three parts: a large centre part 21.8m wide with a length of 10.8m and weighing 200 tonne and two smaller outer parts which were fitted once the central part was bolted up, to give the full deck width of 33m.

On each of my visits I was shown around the partly erected structure by Anthony (Tony) Freeman who was the RE for the bridge. Freeman was the grandson of Ralph Freeman, the founder of Freeman Fox & Partners (in fact, his first name was Ralph, although he preferred to be known as Tony). Although FFP now had a role in the project, he did not work for them at that time and was engaged on a freelance basis, having come from some other large bridge project which had just been completed. He was almost fanatical about large steel bridges and moved around the world to work on them; as soon as one finished, he was off to the next. I found this single-mindedness about large bridges somewhat off-putting as he never spoke of anything else, although my acquaintance with him was short and he might have revealed other interests when one got to know him. That apart, he was the perfect RE for the main bridge; deeply immersed in what was going on and fully understanding all the processes involved. The success of the erection process probably owed a lot to Tony

Freeman's presence. One thing I was not too keen on, however, was his total disregard for safety when he was on the bridge—dragging me after him to show me the most interesting features—even if that meant balancing on narrow access beams 50m above the river.

The last time I saw Tony Freeman was when the bridge was virtually complete, and he was designing a temporary damper system for the stay cables. These had been seen to oscillate in relatively low wind speeds when the vortex shedding frequency of the stay cables in the wind flow was close to the natural frequency of the stays and resonance in the stays arose. The prototype dampers he devised were knocked up in a local garage and worked perfectly until they could be replaced by special tuned dampers for the long term. When he left, he went off to be RE on a large steel bridge over the River Tagus in Portugal. He was in a construction lift with others on that bridge when the hoisting system failed and the lift plunged from near the top of one of the bridge pylons, to hit the ground 100m below, killing all the occupants except for Tony who was taken to hospital in a coma. He never regained consciousness and died of his injuries six weeks later.

Bhutan

I had been chasing work at the Asian Development Bank HQ in Manila and made several visits from Hong Kong to see the engineer responsible for highways and bridges projects which ADB which were financing in the region. Each time when I telephoned the engineer to arrange a date for a meeting, he would ask me to bring him some obscure item of audio equipment which he said could only be obtained in Hong Kong and I would then have to search around the shops in Nathan Road till I found what he wanted. I was never paid for what I brought him, but I hoped to show willing and get some work from him. Finally, on a visit in 1986 he said he might have something and early in January 1987 I got a letter from him inviting us to bid for the design of 12 bridges in Bhutan. It appeared there were just three other firms on the bid list named in the letter.

Getting to Bhutan to see the bridge sites was not easy. Bhutan had no consular facilities in Hong Kong and first you had to fly to Calcutta where there was a Bhutanese Embassy which issued visas, and then take a flight to Thimphu, the capital.

At the time only 3,000 visitors a year were allowed into Bhutan, and it was necessary to get an invitation from the King before a visa could be issued. I got some help from the British Deputy High Commissioner in Calcutta and was promised a visa for Bhutan which I could collect on arrival in Calcutta. Of course, I also needed a visa for India, but getting that in Hong Kong was relatively simple.

It took the best part of a month to make these arrangements and it was early February when I finally set off. I had booked into the Grand Hotel on Chowringhee Road, which had recently been bought by the Oberoi Group, although local taxi drivers knew it only as the Grand. I arrived on a Thursday which turned out not to be a good day for two reasons. First, as I was being shown to my room by a porter, the floor boy approached me and asked if I would like him to get me some beer. I said not to bother as I could get some in the bar later and thought he looked at me strangely. I was to find out why when I ordered a beer in the bar and was told that no alcohol could be sold anywhere in West Bengal on a Thursday. There was nothing but coke and other soft drinks available. Secondly, when I went around to the Bhutanese Embassy on the Friday morning, it was to find that the necessary permission from the King had not yet materialised so that a visa could not be issued. As the weekend was coming up it would be Monday at the earliest before anything could be done. I had to kick my heels for three days doing the sights of Calcutta on my own which involved mostly walking around the Maidan just across Chowringhee Road from the hotel watching the many cricket matches being played there with hordes of beggars following me wherever I went.

The Grand Hotel had a very pleasant swimming pool, however, and I spent some time that weekend in a sun lounger by the pool which was set in a lush garden area. I think it was on the Sunday afternoon when my sunbathing was interrupted by a commotion outside the hotel on nearby Chowringhee Road. There was a lot of shouting and banging of drums and I thought some sort of riot had started right next to the hotel. Ted Foster had told me of the riots in Calcutta in 1946/47 when he was stationed there as a second lieutenant in the army in the lead up to Indian independence. Maybe Calcutta was still prone riots from time to time. But no, it was simply a crowd of supporters welcoming the Indian cricket team which was due to play a Test in Calcutta with the players staying in the Grand. The Test party had just arrived by bus and the players were going into the main entrance to the hotel through an adoring throng.

I finally got my visa and went to Bhutan on a twelve-seater aircraft, flying up the course of the Brahmaputra River. We touched down at the airport around lunchtime and were bussed to Thimphu which was about 15km away surrounded by mountains. I was booked into the only hotel of any note which was close to the *Wang Chu*. It was a two-storey wooden building, reasonably comfortable, but the rooms became very cold at night when the outside air temperature fell below freezing. In the afternoon, I took a walk up the main street which was on quite a steep gradient. Thimphu is at around 2,500 metres and for the first time since my arrival, I felt the effects of the altitude and stopped at a little tea house near the top to get my breath back. The next day I set about hiring a 4-wheel drive Toyota Land Cruiser with a driver to get to the bridge sites and while I was in the car hire office, I met one of our competitors, a Japanese engineer from a company in Tokyo. We got on well and agreed to share the cost of the vehicle and driver, visiting all the sites together and comparing notes as we did so.

Bhutan is nestled against the Himalayan wall to the north, an unbroken rampart of white beyond which lies Tibet. India is on the other three sides: on the west, Sikkim; the east, Arunachal Pradesh and Assam; and the south, Bengal. The country and is divided by four major glacier-fed rivers running north to south with many tributaries. To get from one river valley to another it is necessary to cross mountain passes over 4,000 metres high. The lower slopes of the mountains are heavily forested with pine, larch, juniper, cypress and cedar. The mountains reach over 7,000 metres in the north with snow leopard, tiger and bear still found on them. The valleys teem with birds and butterflies and the rivers are full of trout. For the most part the climate in the valleys is temperate, although quite cold in winter as I found out, except in the low plains in the south which are sub-tropical. Undoubtedly, Bhutan is a country of outstanding natural beauty.

I knew that the State religion in Bhutan was Buddhism, but as I was used to Thailand where secularity runs parallel with religion, I had not realised just how much more deeply religious the Bhutanese were. Religion in Bhutan was not oppressive in any way, however, and the impression I got was of a happy people, living a simple life, without any worry or scurry and secure in their Buddhist beliefs.

The form of Buddhism they practise incorporates numerous female goddesses, *Shaktis*, who, along with Lord Buddha guide Buddhists to the path of salvation and take shape in many mythical forms. As with Buddhists elsewhere, The Bhutanese believe human deeds on earth are fateful as they determine the manner of rebirth in another life. Not surprisingly, with this background, crime is virtually unknown in Bhutan.

Most of the people staying in the hotel had come up from India and were involved with agriculture in some way, either selling agricultural supplies or servicing equipment. Some had been coming for years and knew the country well. They told me that, apart from religion, or perhaps as a result of their form of Buddhism, the one thing that distinguishes the Bhutanese more than anything else is their attitude to sex—they have no hang ups one might say. Marriages happen quietly after a period of living together and divorces are easily arranged. It was even said that polygamy was permissible if it had the wife's approval. Human reproductive organs are venerated as religious symbols and female sex organs are often painted on the walls of houses, not as graffiti, but as artwork. Certainly, on my brief travels I saw plenty happy flirting between the sexes in the villages and settlements we passed through, unlike the oppressive attitudes at the opposite end of the Himalayas in Pakistan and Afghanistan.

Government administration in Bhutan is largely devolved to fifteen districts centred on a *Dzong* (literally meaning fortress) in each district. But management of the road network and certain national administrative functions were undertaken from a central government building in Thimphu which I visited to discuss the project with the engineer responsible for roads. He had been seconded from the PWD in India and he told me a network of more than 1,000km of roads had been constructed which was currently being extended to outlying settlements. Some of the bridges in the project would be replacements for Bailey bridges on the existing network which were all narrow single lane structures (I had come over one of them on the bus from the airport). Other bridges on connections to settlements would be new structures. This was to be Phase 1 of the project and the roads and bridges in the project were all in the western half of Bhutan; roads in the east would be dealt with in Phase 2. He gave me a map identifying all the locations proposed for new and replacement bridges. Another thing I learned from him was that maintenance of the roads and bridges was undertaken by an engineer corps from the Indian army. Clearly India had a strategic interest in Bhutan and involvement in the road system would facilitate rapid mobilisation should there be any military incursions from China through Tibet.

Over the course of the next few days my Japanese friend and I visited each bridge location in our Land Cruiser. We would leave as early as possible each morning taking a packed lunch from the hotel with us. On the first day we remained in the valley of the *Wang Chu* going south to a large multi-span Bailey bridge where the river widened out on its way to join the Brahmaputra and then north to take in two new bridge locations on road spurs to villages. The following days we headed east, crossing two mountain passes to the valleys of the *Mo Chu* and *Tongsa Chu*. The passes were both at an altitude of around 4,000 metres and close to the snowline. The air was thin and bitterly cold, but it was exhilarating to get out of the warm vehicle for a minute or two with prayer flags and tinkling bells all round, looking at the white wall of the Himalayas to the north. If not quite on the roof of the world we were certainly close to it. After locating and inspecting the bridge sites in the second valley we stopped at the *Tongsa Dzong*. This massive fortress was the ancestral seat of the Royal Family of Bhutan and contained the famous *Lama* Centre with twenty temples surrounded by a moat. There was a sign outside in English welcoming visitors and we walked into the main courtyard. A monk appeared and asked in almost perfect English if we would like him to show us round. There were some exquisite sculptures which even I could appreciate, although I could not identify the material from which they were made. The monk laughed—they were crafted from rhino-horn—obtained in the sub-tropical south of Bhutan. So much for the Lord Buddha's condemnation of harming animals.

We generally aimed to be back in Thimphu by the middle of the afternoon in order to write up our perceived requirements for each bridge while they were fresh in the mind, but one afternoon was spent at an opening ceremony for a footbridge on the *Mo Chu* in an area of orchards and wheat fields. This was pure chance as we were on our way back to Thimphu after inspection of a bridge site further up the river. We stopped to let a little procession file off the road and on to the footbridge. It was led by monks beating gongs and drums and we followed them to have a look at the footbridge, which was a rather elegant little suspension bridge, built by villagers themselves and paid for by the local *Dzong*. The chief administrator of the district, the *Dzongda* was the guest of honour. He saw us and insisted we join him at a table set for lunch under an ornate canopy on the bank of the river. I was on his right hand and my Japanese friend on his left. He spoke good English and told me he had studied agricultural science at a college in Hampshire. Once the monks had completed their prayers food was served. It consisted of yak stew and some sloppy vegetables in a vaguely Chinese style, although not exactly up to Hong Kong standards, with sweet cakes to follow. However, it was livened up by local beer and small cups of *Chung*, a fiery spirit distilled from barley or millet served with cups of salted butter-tea. After a few rounds of *Chung*, we were all the best of friends.

When we had inspected all the proposed bridge sites I flew down to Calcutta and checked back into the Grand. Again, it was a Thursday, but now I knew the drill and I had the floor boy bring me some bottles of Kingfisher. I spent the next day at the pool, completing my notes on the bridges and was back at Kai Tak to be picked up by my wife on the Saturday afternoon.

If we were to take on the project, it was clear to me that we would need to find an engineer who was sufficiently experienced and resourceful to be based in Bhutan for a sufficient length of time to survey the sites and prepare preliminary and detailed design of the bridges. There were parallels with what I had done in Nigeria and Sabah, but in this case the engineer would be on his own with nobody else from the firm as back up and would have to hire a local person to help him with the surveys, much as Ibrahim had done for me in Nigeria. I could have done it myself, but I had too many commitments to leave Hong Kong for the time that would be needed. While I had graduate engineers working for me in Hong Kong on bridge design, I did not think any of them could take on such a roll. For the financial part of the proposal, therefore, all I could do was to work out typical costs for a suitable person and put in CVs of two recently qualified engineers in the London office.

The technical part of the proposal was easier to deal with. I had a pretty good idea of the types of bridges I would use and was able to put together an outline programme for survey and design, including basic hydrological studies for flood flows in the rivers. The proposal was sent off to the ADB within the required submission date, but I was not particularly hopeful. In the event, there was some delay in assessing the proposals and I heard nothing for several months. My usual contact in the ADB in Manila had gone on leave and his deputy was not forthcoming when I phoned. Eventually, I got a letter from the ADB saying we had not been successful; the work had gone to some Australian firm I had never heard of who had not been in the list of consultants in the ADB letter. Later I learned the Australian government had made a special development grant to Bhutan outside of the normal ADB lending process, part of which was to be spent on road schemes. It seemed there had been some dirty work at the crossroads.

My promotional activities out of Hong Kong on the Chinese mainland, in Sarawak and in Bhutan had not been very profitable. If we had had greater resources, I would have persisted and I expect major work would have been forthcoming eventually. However, we had insufficient staff for this and from then on, we focused exclusively on Hong Kong.

Kings Road and Farewell to Hong Kong

The Hong Kong office was running well. Ted Foster had done the groundwork and Tony Bowley and I had consolidated the firm's position as a middle ranking civil engineering consultant. It was unlikely we would hit the heights we might have reached if things had been played differently at the start. We had missed the boat for the mega projects which the government had in the pipeline and large British consultants such as Atkins and Motts were building up staff in anticipation of competing for these against the established big two of Maunsell and Scott Wilson. However, we were well regarded for bread-and-butter work with Highways Department on one hand and Housing Department on the other. Tony had just obtained a project for reconstruction of Kings Road on Hong Kong Island, from Causeway Bay to North Point, later extended to Quarry Bay and I had started on a new site formation job for HD with CHP on the steep hillside above Tsuen Wan. The housing block repair supervision was also running well with regular fee income from it. In addition, I had picked up work with Caltex who were in the process of moving their oil terminal from Tsuen Wan to an old oil terminal site on Tsing Yi Island.

We had a about twenty people in the office and another twenty or so on supervision of projects in various sites around the colony with Tony and me the only *Gweilos*. Most of the staff had been with us for years and were very loyal and happy to work long hours if the occasion demanded it. We organised regular junk trips for the whole office, snooker evenings and parties for the staff, both from the office and the sites. A weekly highlight in the office was Friday lunch where we would go to a restaurant chosen by the staff, and with the variety and sheer number of regional Chinese restaurants in Hong Kong that we might go for a year without visiting the same place twice. The staff would order the food, such as snake in the autumn when it is said to fortify one for the coming winter or spicy Szechuan duck with steamed bread. In those days everyone simply dug into the various dishes with their own chopsticks, unlike now after the SARS scare in 2002 when separate sets of chopsticks are provided for serving and eating. These lunches gave me some skill in using chopsticks and I was said to be able to "eat like a Chinese" as my wife observed, not necessarily being complimentary. It is sometimes said of Hong Kong that although the British and Chinese forged a successful working partnership in the colony, the separate races did not really like each other. There may be some truth in that in general, but it was certainly not the case in our office, where triumphs or misfortunes were felt and shared by Chinese and *Gweilos* alike, and we got along together very well indeed. I suspect the same applied in other enterprises working for a common goal.

Another highlight for me, outside of the office this time, was the annual Glasgow High School dinner, a men-only affair, which was held in the week before the Hong Kong Sevens tournament. The dinner was organised each year by Glen Docherty who at that time was also chairman of the committee organising the Hong Kong Sevens, before it became a major international event. Glen must have been around five years older than me and had played in the First Fifteen when I was in the first year at the school. The dinner was always held in the Hong Kong Club of which Glen was a member (he went on to be president of the club) and we could usually muster around twenty former pupils, more than any other UK school, even Eton, or so we said. As the dinner was just before the Sevens, we often invited guests who had come over to watch the rugby and one year I was seated next to John Rutherford who many considered to be the best fly half to play for Scotland. He had just retired from international rugby after suffering a cruciate ligament injury. They were rather boozy dinners and after the meal, which was usually excellent, the port would circulate for a long time.

We were well settled in our new house in Long Keng where we were friendly with our neighbours Graeme and Jane Hagen, who lived below us in a terrace of village houses at the edge of the old paddies. At weekends we would get together for barbeques or just beers on our veranda or theirs (once the sun was over the yardarm as we would say). Graeme was a barrister much of whose income came from defending triad members who had been caught in some illegal and often vicious activity. His forte was not so much having his clients acquitted, but in successfully pleading mitigation and getting reduced sentences for them, perhaps 2 years instead of 5. As he put it to me, he did not think he had ever defended an innocent man. Graeme was something of an expert on the organisation of the triads and knew the hierarchy from the Lodge-Masters at the top, through the executive Red Staffs and White Fans down to the humble Straw Sandals and Pole Carriers. Graeme and Jane had two daughters, both in their teens, who were at school in England and who came out to Hong Kong in the holidays. Another lawyer, Bill Radcliffe lived next door to Jane and Graeme with his wife. Their son, Edward, was a year younger than Stephen and the two boys would run around the old paddies seemingly unaffected by the ever-present mosquitos. For the children it was in many ways an idyllic existence. They could play football or badminton in our garden, and we often walked in the country park out to the beautiful beaches of Tai Long Wan to the north-east, sometimes along with Graeme and Jane and their dogs, passing through small Chinese villages with fading paper door gods at the entrances of the old buildings. There was rugby with Kowloon minis every Sunday for David and John at training pitches near Kai Tak Airport and regular tournaments against Football Club in Happy Valley and teams from the army garrisons in Fanling and Stanley. There was also the KCC with its squash, tennis and swimming. The only problem was to find time to fit in all the activities available to us. For children, Hong Kong was a very safe place when we were there and while there was always a background of triad violence, it never spilled over to the expatriate communities.

A sad event had occurred in 1987, however, as Tony's wife Sue died of cancer of the colon. She had had surgery the year before, but the cancer had returned. Tony and Sue had many friends and the funeral service in St Andrews Church in Nathan Road was packed. Tony and I along with the project engineer from the NTDD, Rob Dickie, had written a paper on the Trunk Road for publication in the ICE

Proceedings and the paper was to be formally presented at an ICE meeting in London in October 1987. Tony flew over to give the presentation and while he was in the UK, he was able to visit some of the members of Sue's family who had been unable to get to Hong Kong for the funeral. The trip to the UK was therefore of some benefit in keeping him busy in the immediate aftermath of Sue's death.

The Kings Road reconstruction project which was now starting up was largely highways oriented, involving complicated traffic management procedures on the busy route, especially when the Quarry Bay tram tracks were realigned on a new formation. I had two areas of involvement: geotechnical engineering where road widening required slopes to be steepened, and the design of several new footbridges over the road.

By now I was more familiar with the geotechnical issues peculiar to Hong Kong and no longer needed specialist help in dealing with them. Most of the slopes along the road were stable and well protected against water ingress by *chunam* or concrete on the surface of the slopes, but there were some rock cuttings where the stability of the cut slopes needed to be investigated for the potential failure modes described in the Kai Liu site formation project. There were also more weathered slopes where caisson walls had to be designed to retain the slope and a particular location where a high masonry retaining wall overlooked the road. Fortunately, the wall did not have to be disturbed as the road was not being realigned at that location, but there was some concern as to its long-term stability. The wall dated back around 100 years to the 1880s and had been built as the original developments in Central on Hong Kong Island were expanding eastwards past Causeway Bay. I was all for leaving it without any special strengthening measures as it looked in sound condition and discussed this with the GCO. We ran various computer models on its stability which were available at that time, but these were inconclusive, except that it could be demonstrated that its factor of safety against failure with the highest possible ground water level must have been at least 1.0, otherwise it would have failed. Nobody would design a new wall with such a low safety factor, but I argued as the wall had proved itself over 100 years, this would give a reference point. Back calculation with a safety factor of 1.0 was then carried out to derive properties of the soils at its base in line with that factor. I proposed positive drainage measures behind the wall to ensure the ground water would always be much lower than the high level used in the analysis and then reran the models with the derived soil properties which now showed the wall to have a much higher factor of safety. The GCO accepted this, and the wall was left untouched except for the improved drainage. Since then, I have used this simplified procedure on several other old walls. In 2015 I had occasion to visit a contractor's office on Kings Road and was pleased to see the wall was just as we had left it in the 1980s.

As the famous Hong Kong Island trams ran down the centre of the road with a notional two traffic lanes on either side of the tramway, the footbridges had to have clear spans of around 45 metres from one side of the road to the other without intermediate supports as well as having to carry the overhead conductors for the trams. They were to be provided with canopies for shelter (otherwise pedestrians would not use them in heavy rain and would run across the road through the traffic). Nowadays all footbridges in urban areas in Hong Kong have canopies, but then it was a new requirement. They also were to have a wide footway as high pedestrian flows were expected. With all these features, the dead and live loadings were heavier than usual for footbridges and there was little scope to produce slender elegant structures; steel trusses of one form or another were called for. I have always liked steel truss bridges myself from experience of them at Ballachulish and Kinabatangan and especially when they are of Warren girder form which was how I designed those. I was one of the first engineers to do this in Hong Kong and I am pleased to see virtually all footbridges on Hong Kong Island and Kowloon are now in the form of Warren girder trusses.

One evening in early July 1987 Peter Fraenkel telephoned me at home. I was playing squash when he phoned, and he asked my wife if I would me call him back. When I got home I did so and without much in the way of small talk he invited me to become a director of a new company he had formed, Peter Fraenkel International Ltd (PFI) to which most of the work of the partnership was being transferred. I would be responsible for all bridges, including those on highway schemes in England for the Department of Transport which he said the firm was now obtaining on a regular basis. The job would be in London, and he would like me to take up the appointment in around one year's time in the summer of 1988 which would give time to make new arrangements for Hong Kong. I said it would certainly be of interest which he took to be a yes.

On the face of it this was a good opportunity to take a senior management role in an international firm of engineering consultants. It was the sort of thing most people aspire to and to be offered it at the age of 43 could be considered good going in a profession where promotion to such roles in a London firm could be problematical. It was also something that was difficult to turn down and at the same time remain with the company. While I was happy in Hong Kong and things were going well, it would be a step up if it worked out. But that was an issue that bothered me—would it? I had been out of the UK scene for most of the last decade and my experience of the rampant inflation and industrial unrest in Britain of the 1970s had not been good, one reason so many of us were happy to be in Hong Kong away from such things. However, it was said things had changed since Mrs Thatcher had put the trade unions in their place, so perhaps I should give it a go. For better or worse I agreed.

Around three months later we had a meal with Graeme and Jane in a seafood restaurant on the Sai Kung waterfront. It was the evening of the Mid-Autumn (or Harvest) Festival, the fifteenth day of the eighth month of the Chinese lunar calendar and a feast of the full moon. The waterfront at Sai Kung was a particularly propitious place to be for the festival as the full moon of autumn rises over Sharp Island to the east. Although it was still early evening when we got there, the moon had already risen, greeted with firecrackers and gongs from boats at the quay and there were many tables in the open air in front of the restaurant, mostly taken by Chinese family groups, and a few by *Gweilos* like us. However, the owner knew us and a large table for four adults and three boys was quickly found along with seven chairs. Graeme and I went over to the fish tanks to make our selection, a reddish-brown grouper of suitable size for our party, some crabs and a catty and a half of shrimps. There was no need to specify how the food was to be cooked as we knew it would be delicious: the grouper would be steamed with soy sauce and ginger and a splash of hot oil to make it sizzle as it was brought to the table, the crabs would be stewed in a spicy sauce and the shrimps would be stir fried in the wok with garlic and a dash of rice wine, all served with mounds of steamed rice. Large bottles of Carlsberg were ordered with coke for the boys, and we sat back to enjoy the company and the vibrant scene around us with happy diners and the usual loud buzz of conversation. As I looked at this scene, I was beset by doubts on the wisdom of my agreement to return to the UK and work in London. How could I give up living in such a delightful place as Hong Kong? A little voice inside me was telling me that things would not go well back there—and there was still time to back out before too many boats were burned. But after a few beers I could convince myself that all would be well. Moreover, the move was still nine months away and much could happen before then. But I should have paid more attention to the little voice and taken the advice; these little inner voices are rarely wrong.

Over the year we prepared to go back to the UK in July 1988. This included arranging schooling and we got David into Glasgow High School, my old school, and John and Stephen into Belmont in Whitecraigs, quite close to our house in Eaglesham which was currently let. My jobs in Hong Kong were all going well as were Tony's, but I continued to have forebodings and concerns on what I was giving up and more so on hearing on the grapevine of cash flow issues in London. At the same time, the firm was in the news with the opening of the Rama IX bridge in Bangkok in January 1988 which I attended, as well as presenting a paper on the design of the approaches at a conference on large bridges timed to coincide with the opening. I made some visits to Indonesia where we were winning marine works for the port authority PT Pelabuhan II, through an agent, Freddy de Lima who reported to me. These things gave me some confidence it would work out.

In the lead up to our departure we had a memorable farewell dinner in the KCC, and I paid a fee of three times the monthly subscription to become an absent member—the best investment I ever made. Tony organised a family junk trip for all the whole staff to Lamma Island and in hot summer weather we swam off the junk in a bay on the south coast of the island before eating in one of the fish restaurants in Sok Kwu Wan. Finally, Ah Yeung was paid off with a bonus, our furniture was crated up and loaded in containers for shipping to the UK and on 15 July 1988, in a violent thunderstorm, we closed the door of the house in Long Keng for the last time. We stayed that night in the New World Hotel in Tsim Sha Tsui and left Hong Kong the next day, heading for KL and Kuantan for a holiday before going on to the UK at the end of the month, for me to start in the London office in August. As with Caesar's crossing of the Rubicon, *alea iacta est*.

Part 2

London

> The Fathers of the City,
> They sat all night and day,
> For every hour some horseman came
> With tidings of dismay.
> *From Horatius, Lord Macauley*

It took only a few days in the London office in that first week in August 1988 for me to realise I had made a serious mistake in leaving Hong Kong and looking back over all the years since my view has not changed. Things were much worse than I would have thought possible. All projects were now undertaken by the new company with the grandiose title of Peter Fraenkel International Limited. There was plenty work in the London office, largely relating to highway improvement schemes throughout England, but for reasons which soon became apparent the firm was in dire straits financially with a mounting overdraft at Lloyds Bank. Management was appalling with a bloated board structure, a loose accounting system which failed to properly recover costs and questionable technical competency in many of the staff. It had been over ten years since I had last been in London and during that time the vibrant optimism of the firm had gone, to be replaced by a downbeat feeling accentuated by the worsening cash situation. The contrast with our happy, profitable office in Hong Kong could not be more marked. How had things gone so badly wrong?

There seemed to be several factors as there usually are in these situations. First, the entrepreneurial people at the top in the 1970s and the well-connected partners such as Roger Postlethwaite and John Stanbury had gone along with the competent engineers who worked under them. They had been replaced by a board of yes-men with little flare or technical expertise. Everything was decided by Peter Fraenkel as chairman without question. Secondly, although there were substantial highway schemes on the go, they were not run efficiently. They had been obtained without much thought on whether they could make a profit, they were over-staffed and some of the people working on them were very inexperienced resulting in much abortive work to correct their mistakes. Thirdly, the firm was paying over the odds for offices, as in addition to an expensive office in Wimbledon there was an office in Whiskin Street not far from Angel Underground station for work not connected with roads.

Substantial losses had arisen from an ill-judged attempt to set up an office in the French-speaking part of Cameroon in West Africa. Some jobs had been obtained for which fees were paid into a bank account in Cameroon, but the local partner in the venture, an African lady, had access to the account and emptied it, then disappeared. To give a local partner access to an account without requiring other signatures was naïve and something anyone with a knowledge of West Africa would never have contemplated.

As the financial position worsened with the bank overdraft approaching £300,000, Lloyds demanded personal guarantees from the directors, and I had to provide one to match my shareholding in the company. Just when it looked as if the company would go under, an approach was made for a takeover by British Maritime Technology, a recently privatised government organisation formed by a merger of the British Ship Research Association and the National Maritime Institute. At that time, their field of operation was mostly research in ship systems design and vessel handling in harbours, and they saw Fraenkel's civil engineering capability as potentially useful in being able to offer a wider service to clients in the marine sector. Protracted negotiations followed, characterised by involvement of expensive legal and accountancy specialists. BMT's original idea had been to take over the Fraenkel organisation lock, stock and barrel, including roads and bridges, even though their real interest lay in the ports and harbours expertise which the firm still possessed. This would have been the best outcome for most of us, but it did not suit Peter Fraenkel who would have had to accept an overall loss of control. What was eventually agreed, therefore, was to set up a joint venture between Peter Fraenkel and BMT called Peter Fraenkel BMT Ltd for the maritime business and the work in Hong Kong. The part of the firm dealing with UK roads and bridges along with general civil engineering was to remain outside of the deal to operate again as PFP.

BMT paid a substantial sum to acquire 50% of the ports and harbours business which cleared the most pressing debts and enabled payment of fees of over £100,000 to the expensive advisers. I had my guarantee released, although it took months for Lloyds Bank to provide me with confirmation of this, but I got nothing else out of it other than a position as partner in a completely new partnership formed to run the roads and bridges business.

Now there were just three partners, Peter Fraenkel, Colin Campbell, who ran the roads projects and me. There were several bridges on each of these projects for which I had overall responsibility. However, I had an excellent bridge engineer called Alf Treserden, a bachelor whose hobby was playing the stock market, reasonably successfully as far as I could make out. I did little bridge design myself leaving this to Alf and concentrating on management of the business and promotional activities. In addition, I took on some work back from the new Fraenkel BMT joint venture on a sub-consultant basis, mostly relating to a port project we had been doing in Indonesia which had gone into the BMT part of the business. This was conversion of a general cargo quay to a container terminal in Tanjung Priok, the port of Jakarta, set up for us by my old friend the agent Freddy de Lima. On two occasions, while the regular RE was on leave, I filled in as RE on the contract in the port for periods of one month at a time. I did other things for Peter Fraenkel BMT such as carry out basic design checks where the joint venture company lacked the necessary expertise. I also returned to Hong Kong to complete a job I had obtained before I left. This was the design of three bridges for the cargo terminal in Kai Tak airport which was still in operation at that time while the new airport at Chek Lap Kok in Lantau was under construction. I made other visits to Hong Kong over the next year or two, but these became less frequent as roads and bridges were of less interest to the BMT side of things, now run in Hong Kong by Peter French, and very sadly Tony Bowley developed an inoperable brain tumour and died back in the UK after a short illness.

By now I had been elected a Fellow of both ICE and IStructE and at first sight the partnership was in a good position, debt free itself with a substantial workload and without the burden of the expensive, non-productive board. But even here there was a serious underlying problem; most of the road jobs we were doing had been obtained at such low fees it was impossible to run them profitably. Bidding for highway jobs at low prices had been a deliberate policy, aimed at increasing the firm's profile and building up the highways workload to impress clients. In certain circumstances "buying" work in this way could pay off so long as there were profitable jobs to balance the losses. Unfortunately, that was not the case as none of the jobs which had been obtained were profitable. As intended by the Department of Transport (soon to become the Highways Agency) they were also long-term appointments, and we would have to put up with the low fees for years. The situation was being made even worse by the mediocrity of many of the road engineers who had been recruited into the organisation who were unable to perform efficiently. There had been a complete lack of commercial awareness in setting up this UK roads business and now I was sucked into it as a partner. In addition, unforeseen liabilities in PFI not dealt with in the sale to BMT emerged. These liabilities ultimately had to be covered by PFP.

The regional director of the new Highways Agency in the East Midlands where we had some ongoing schemes pressed Peter Fraenkel to open an office in his area on the promise of further work to come. Accordingly, we opened an office in Leicester and some more jobs were indeed forthcoming. I took over responsibility for the office with a competent highway engineer called Keith Addison working under me to run the roads work and look after the place on a day-to-day basis. I also shifted all the bridge work to the new office and recruited bridge engineers from the Leicester area to form an effective bridge design section. Alf Treserden did not want to move out of London, and he remained in the Wimbledon office for a while in a part-time capacity which suited him as he could devote more time to his stock market investments. With the firm being run on a more rational basis, things gradually improved, although we were still not profitable overall. Then in 1995 two events changed the situation radically.

The first was a rethink in UK government policy on road building. In the early 1990s the UK was investing a lot of money in the road network, but the growing environmental lobby was pressing for the road building programme to be abandoned, or at least drastically curtailed. With an election coming up, the weak Conservative government led by John Major, under attack from a resurgent Labour Party under Tony Blair, bowed to the pressure and many schemes were cancelled and others reviewed. In the space of a year, we went from running ten road schemes in England to just three. The new Leicester office lost three schemes itself in that year and was left with only one scheme which was for a major new junction on the M40. Two years later Labour came to power and killed off all the schemes which were left. With no help from the ineffectual Association of Consulting Engineers or the ICE, some of the consulting engineers who had lost a lot of their income, PFP included, banded together to raise a legal action against the Highways Agency, but despite incurring substantial legal fees this got nowhere and was abandoned. Of course, five years later, in the early 2000s, major roads in the UK became so congested that the government decision to stop road building had to be reversed, but by that time we were out of road design for good. This had a severe effect on our revenues and debts piled up again while we ran down the staff numbers and tried to cut office costs. We kept the office in Leicester as we had obtained other work there to replace the cancelled road schemes, but we moved out of the office in Wimbledon to temporary accommodation near New Maldon and then down to Dorking in Surrey. However, as it turned out, being forced out of the roads business in this way eventually proved to be a blessing in disguise; we were shot of all the unprofitable jobs.

The second event was more positive, especially in the light of the first. BMT came to us and asked if we could buy out their share in the UK business of the Peter Fraenkel BMT joint venture. Their price was one pound, with the proviso they could take over the Hong Kong business for themselves and Peter Fraenkel's residual interest in it would cease. BMT had made a big mistake when they bought out the PFI ports business: instead of running it themselves or recruiting competent staff to run it they kept on the PFI directors who had made such a mess of things. They were not to know that these people were so incompetent, except that they might have wondered how it was that PFI were in such dire straits when most other UK consulting engineers were doing very nicely at that time. Management of the business in the UK since they took over in 1990 had been even worse than PFI's and they had lost many of our old clients, especially those in London, partly as a result of having moved their office base out of London to Southampton where they shared an office with another BMT company. They were losing several hundred thousand pounds a year and wanted to cut their losses and get out of the UK civil engineering as soon as possible. I agreed with Peter Fraenkel we should go for it and a simple sale agreement was drawn up without involvement of any lawyers or accountants this time. We changed the name immediately to Peter Fraenkel Maritime Limited and registered this at Companies House. I became managing director of the new company. To some extent I was worried about what we were taking over, but there were still a few jobs on the books, and I felt we could build back up again on the strength of the old name. As far as Hong Kong was concerned, I was sad to lose the connection, but I had not been involved there since Tony died and with the character of the office changed, I accepted its loss. In the event BMT made little of it in Hong Kong without Tony and eventually sold that business as well. To be fair to BMT they became very successful once they reverted to the high value specialist projects they understood, especially when the one PFI director who was competent, Peter French, rose to become their CEO.

I received repayment for my old PFI shares which had remained in the Peter Fraenkel BMT joint venture and I invested this in the new company we had set up. Peter Fraenkel put in more money to give us adequate working capital and an overdraft facility with Barclays Bank was easily arranged without the need for any guarantees. We had an office in Southampton with some staff and we took on more as work increased, but in fact, only one of the original Southampton staff remained with us after six months. This was Guy Olliver, a specialist in the marine physical environment such as waves, currents and sedimentation, who was to play a big role with us over the next fifteen years as our ports business built up. I was now looking after offices in both Leicester and Southampton as well as trying to expand the new business and was becoming over-stretched until we closed the Southampton office and moved the ports business we had acquired to Dorking, not quite back to London, but not far away.

During the years I was overseas a separate partnership had been formed in Scotland, combining the old business in Glasgow with an Edinburgh firm specialising in water supply projects called Leslie & Reid and the firm in Scotland now operated as Peter Fraenkel Leslie & Reid. Ken Archibald was the partner in the Glasgow office with Willie Roxburgh as associate (later partner). The partner in the Edinburgh office was David Neale. The firm in Scotland was well run in a quiet way and quite profitable. However, despite the opposition of Peter Fraenkel who was also a partner in the Scottish business, it ultimately merged with a much larger firm of consulting engineers from Leeds called White Young Green. Following this, WYG made an approach to us to buy out our Leicester office with its work as a going concern and this was agreed, Colin Campbell going over to them with the business.

I was pleased with all this and could now focus on developing PFM in which I had a free hand as Peter Fraenkel had stepped back from management, while retaining a controlling financial interest and continuing to provide finance in the shape of loans which were converted into preference shares held by his family trust.

A lot of promotional effort was now needed if we were to survive. Amongst the major clients lost in the BMT period were Shetland Islands Council and with them consultancy work in the Sullom Voe Terminal along other marine civil engineering work in Shetland. Peterhead Bay Authority and the privatised Port of Tilbury had also gone as clients. They had all been excellent clients to work for and I made strenuous efforts to recover our position with them.

The first one we got back was Shetland Islands Council. We had had had a long association with SIC going back to our role as consulting engineers for the oil terminal jetties in the 1970s and it was difficult to credit that BMT had let that lapse. I went up to Shetland and met Alistair Inkster who was Director of Engineering in the council who knew me from Trondra and Burra Bridges days, and then went on to the terminal at Sullom Voe where I met the Port Director Captain George Sutherland and his port engineer Dave Drummond. I suppose I was lucky in the timing of my visit. It so happened that a firm called Arch Henderson (with offices in Aberdeen and in Shetland itself) had started to work in SVT, but the local partner had left the company abruptly for personal reasons and I was able to step into the gap. In my meetings I stressed to both Alistair and George that the joint venture between Peter Fraenkel BMT was gone and PFP or PFM as it was now styled was back in business on its own. They were both openly critical of the joint venture's performance—deadlines not met, lack of communication and so on—and they wanted nothing to do with the directors and staff who had been involved. Fortunately, they accepted my description of the complete break which had occurred, and both had a great deal of time for our old organisation which had performed so well in the past. Right away we were awarded a commission for a comprehensive assessment of the condition of the four jetties in the Sullom Voe terminal which got us started, to be followed by responsibility for annual maintenance contracts and much more besides, including design of new inter-island ferry terminals. For the maintenance work I was greatly assisted by Robin Theobald who had re-joined us shortly after the setting up of PFM. Robin acted as RE on the contracts for over twenty years. This provided us with a good background income from time-charge fees for Robin's involvement on site and in the office. He came to know the jetties as well as anyone in SIC who were happy for him to take over responsibility for virtually all maintenance planning and supervision.

In the summer of 1997, I visited Peterhead to meet the Chief Executive, of the Peterhead Bay authority, Captain Bert Flett to try to recover our position there. He knew the firm well having been in a similar role at Sullom Voe in the years just after the terminal opened in 1978, but I had never met him as I had been overseas. He suggested I call back in six months when he might have something for us.

I then went to Malaysia for three months where Patrick Augustin had won a project to design a large urban flyover on Jalan Tun Razak in KL. The flyover effectively added separate 3-lane dual carriageways above the existing Jalan Tun Razak road for 1.1km with single large piers to support the carriageways in the central reserve. Patrick's engineers set up a grillage model of the superstructure which was to be of steel composite construction, and I carried out the detailed design myself to BS 5400 working in Patrick's office in PJ and staying in the Commonwealth Club in Bukit Damansara. My wife and youngest son came out for a few weeks to stay with me in the Club. With email now becoming effective for rapid communication, I was able to manage the business from afar and being in Malaysia was something of a break from the travelling around the UK. This was the last major design work I did myself until years later in 2016 when I took on a much bigger design assignment back in Hong Kong.

With this pleasant overseas interlude, it was late February 1998 before I went back to Peterhead for the follow up visit to Bert Flett. Again, we were very lucky in the timing of my second visit. A few years earlier we had designed a jetty for PBA for use by North Sea oil supply vessels. It was called the Princess Royal Jetty as it had been officially opened by Princess Anne. To save money PBA had instructed us to suspend service pipes below the jetty deck, despite our recommendation that all services should be accommodated in a service trench built into the jetty for protection. The inevitable had happened just before I made my visit; a severe storm with winds from the south-east had driven large waves into the harbour and these had carried away many of the service pipes slung under the jetty. We were immediately given the job of sorting things out and this led to many other commissions in Peterhead Harbour over the next fifteen years.

The third client lost by Peter Fraenkel BMT had been the Port of Tilbury. When I had joined the firm in back in 1974, we had a virtual monopoly of all consulting engineering services in Tilbury Docks. Apart from designing the floodgate, we were responsible for a new lead-in jetty at the Western Entrance Lock, we devised and implemented a scheme to close the redundant entrance to the docks at the old tidal basin and organised many repairs to facilities damaged by accidental impacts from vessels. When the PLA was split up and the Port of Tilbury was established as a separate privatised entity in the 1980s, we continued to obtain work from the port engineers, but then, with the move out of London and the joint venture's mismanagement, the work tailed off. I aimed to set it up again, helped once more by Robin Theobald. He had worked for POT for some time before coming back to us in 1995 and had as good a knowledge of Tilbury Docks as he eventually acquired of the Sullom Voe jetties. To start with we advised on repairs to POT's cruise terminal jetty which Robin knew well, and work built up gradually from there until we got a series of heavy-duty paving schemes and studies for long-term dock expansion on the river frontage.

In this way SIC, PBA and POT provided us with steady background income on design and management of maintenance of their facilities as well periodically paying substantial fees for the design of specific projects. In addition, we began to expand our client base generally and our turnover increased year by year. This higher turnover was sustained throughout the next eight years and peaked in 2010, by which time we had reserves of nearly £500,000 and I had thoughts of clearing all the debts which had built up in the 1990s. Everything was now concentrated in a single office in Dorking, and we had excellent control of our finances, greatly aided by our part-time accountant Ted Chenery, who was assisted by my marvellous secretary, Liz Rae. Ted provided clear management accounts at the start of each month, in sharp contrast to the loose accounting system in use when I returned from Hong Kong.

In 2005 we got a further boost when we were joined by Roy Glenton, an experienced marine civil engineer who had been working on his own account for several years. Roy bought into the company with an investment which gave him 30%

of the equity which was half of my holding of 60% (by this time Peter Fraenkel had reduced his holding to 10%, but still had control under the articles of association and by means of 26% of the voting shares). Roy had started his career with Crouch & Hogg in Glasgow and had then worked for various consulting engineers and contractors including Mowlem Marine with whom he had been Chief Engineer. Roy was a very good engineer with a hands-on approach to design matters and he also brought with him good connections with a range of contractors for design and build projects. He lived in Largs on the Firth of Clyde, so we opened a separate office for him close to Glasgow airport. As the firm had expanded again, I had been struggling to manage design work as well as run the business, but with Roy in the company I could leave much of the design management to him. It was a very successful relationship which lasted nine years until 2014 when Roy left to go freelance once more in somewhat unfortunate circumstances.

With Roy's connections complementing my own, we had our best years since the restructuring of the business with a mixture of traditional consulting engineering for long-term clients which I looked after, and design and build work with contractors which was Roy's area of expertise. To broaden our perceived capabilities for a range of clients, in the autumn of 2005 we dropped the name "Maritime" and renamed the company Peter Fraenkel & Partners Limited, effectively back to PFP which was still well known and respected in the industry. Also, in 2005, Peter Fraenkel, by now ninety, stopped coming to the office. I sent him summaries of what was going on and for a year or two and would sometimes get queries from him, but after a time the queries stopped and in November 2009, he died aged 94.

To go with the steady flow of work from our old clients and the new ones which Roy brought to us, I got an introduction to a civil engineering contractor in Portugal with whom we enjoyed a long and fruitful relationship almost up to my retirement. The contractor went by the acronym of SETH—*Sociedade de Empreittades e Trabalhos Hidraulicos*. The business had been established originally by a Danish firm who still maintained an interest and it was run by a Dane, Villy Petersen, one of the best engineers I ever met. SETH were seeking to bid for a design and build contract for a 2km long jetty to be constructed in the port of Kamsar in Guinea and needed a consulting engineer for the design. This was for an entrepreneur in New York who had won a concession from the Guinea government to mine alumina about 100km up country and refine it into pelletised aluminium for export from Kamsar. An American firm of consulting engineers had produced a tender design for bidders to price. Working closely with Villy we produced a new design which was greatly superior to the conforming design and SETH won the contract. We went on to complete the detailed design of this jetty as well as smaller service jetties and other facilities and subsequently worked with China Harbour on a separate alumina jetty in Kamsar.

Then in 2008 more heavy clouds appeared on the horizon. Excessive risk taking by banks and mortgage lenders in the United States precipitated a severe global recession. At first, we were unaffected by it as we had a good workload and none of our clients cancelled ongoing projects. As well as work in Shetland and Tilbury, we had other large projects running including a jetty and breakwater in Peterhead Harbour and the Kamsar jetty, the combined fees from which pushed our net profit to over 20% of turnover in each of the 2009 and 2010 financial years. However, I could see that, as the recession worked its way beyond the banking sector which had caused the problems and into the wider economy, fewer new projects were likely to come forward and it would be best to sell up while we were still in a sound financial position. We had good connections with senior management in Atkins, at the time the largest independent UK consultant. Although they were a large organisation, they were short on marine civil engineering expertise, and Roy and I reasoned that our background in that area of work could give them a ready-made marine capability. I had several meetings with one of their directors which led to definite interest from them and a provisional offer to purchase the business. This involved a fair price based on a multiplier of 5 on the annual profits and what I considered to be generous workout payments for Roy and me to be allocated in the ratio of our shareholdings, provided we remained in the organisation for at least two years from completion. Their offer seemed almost too good to be true: any debts would be cleared off and we would have enough left over for a reasonably comfortable retirement. We agreed a target completion date of 30 November for the acquisition, subject to due diligence. Before we could start the due diligence process, however, we needed to get the agreement of the Fraenkel family who still had ultimate control over any sale under the company's articles of association through a key holding of voting shares. I could see no reason why the offer would not be to their liking as their loans would be repaid in full which might not otherwise be the case and they would also receive a substantial payment for their remaining 10% of ordinary shares in the business. Despite the obvious benefit to them it still took six weeks for approval to be given. We engaged Michael Archer of Beale & Company to agree the sale and purchase agreement. Michael had acted as company secretary since the early days of PFM, and I had come to know him well and trusted him to look after our interests.

The delay in getting agreement from the Fraenkel family put back completion and a new target of the end of January 2011 was set. The due diligence process ran through November and into December and as far as I could see it was going well. Around the middle of December, it was reported in the press that Atkins were to make 300 staff redundant as a result of the recession which was now beginning to bite in construction as clients put off new projects until the economy picked up. I was not unduly concerned at this as I believed Atkins would still want to increase their marine civil engineering capability for strategic reasons. But a few days before Christmas I got a phone call from the director who had been put in charge of negotiations on the takeover. The main board had decided to halt all acquisitions, including ours, for an indefinite period. Price was not an issue, he said, but it was felt it would not be good for employee relations to be making staff redundant while at the same time taking on other staff in a company buyout. He went on to say he had argued the case for the PFP acquisition to go ahead, but he had been overruled. In that telephone call the marvellous deal which had seemed in our grasp faded away. I suddenly felt very tired of it all.

Would it have made any difference if the Fraenkel family had responded in a timely fashion? It is difficult to say. If the due diligence process had been finished in time for completion by the end of November, the sale might well have gone through as planned. On the other hand, the Atkins main board could have been considering staff redundancies for some time before the announcement was made and might have blocked the acquisition in any case. In any event, the collapse of the deal condemned us to further years of financial difficulties as work was now drying up. However, it must be said that the Fraenkel family were generous in writing off a large proportion of the loans when the time came for final settlement of them.

During 2011 we completed a major job in Peterhead Harbour. While we had been working on that job Peterhead Bay Authority had taken over the Harbour Trust, which was responsible for the inner harbour at Peterhead and had become virtually bankrupt for various reasons. The combined authority was now called Peterhead Port Authority. We continued to get work from them, but now only on a relatively small scale. Design of the large Kamsar jetty had also been completed, although we started to prepare designs for two smaller jetties. We were still working on the maintenance contracts at Sullom Voe, but soon we had no other sizeable projects on the go, just a series of small things which took up a lot of time and effort for relatively low fees. Our turnover was therefore well down as were our profits and the reserves we had built up dwindled rapidly.

Then a representative from a Danish company called NIRAS contacted us and asked for a meeting to discuss possible collaboration in the UK. This was Jesper Harder who was a driving force in the company and had responsibility for much of its business development outside of Denmark. One of NIRAS's main interests was offshore wind and they had a contract with the Danish energy company DONG for design and management of a windfarm in the North Sea to be serviced from Grimsby which gave them an introduction to work in the UK. Roy Glenton and I met Jesper Harder with two of his colleagues in Dorking and we got on well. It was clear NIRAS had some interest in a possible takeover of PFP. They got us involved in some of the work in Grimsby and their interest heightened towards the end of 2012 and in March 2013 an offer was made. It was in the form of a merger and nothing like as good as the Atkins offer of 2010, but our position had deteriorated to such an extent that was to be expected. At least it would enable us to continue without sweeping redundancies and the way it was structured there were prospects of workout profits for Roy and me if things went well. Once more I had to go through the process of persuading the Fraenkel family to agree to the terms. They would only get 50% repayment of a loan which had been converted into preference shares and this was a sticking point, but it was stressed that the NIRAS offer was probably the best we could hope for and the alternative to this "haircut" for them might be the firm going into receivership following which they would get nothing.

We engaged Michael Archer of Beale & Company to act for us as he had with Atkins and another due diligence process was carried out. By the summer, the arrangements had been finalised and the merger was completed on 26 July 2013 in Michael Archer's office in London. Once the documents had all been signed Michael brought out the champagne, but the Fraenkel organisation was no longer independent and having been involved in it for so long I did not feel like celebrating. The name survived for some years in NIRAS Fraenkel as the new company was called until NIRAS dropped the Fraenkel part after I retired from the company in 2017. *Sic transit gloria mundi.* In truth the glory such as it was had passed with the rescue by BMT twenty-three years before.

The Other Side of the Coin

Perhaps one day it will give us pleasure to remember even these things.
From Aeneid, 1.203, Virgil

In the 29 years between my return from Hong Kong and my retirement, despite all the turbulence within the organisation, I was responsible for many more civil engineering projects, some large, some small, and some of little interest, but a few quite memorable to me at least. It was sometimes a struggle to focus on completing them in the financial turmoil I had found myself in, but at the same time working on them was often a relief. A few of these projects which I found interesting and had some pleasure in carrying out are described here.

Qandil Bridge

The Iraqi Army under Saddam Hussein invaded Kuwait on 2 August 1990. It was claimed by Saddam that Iraq had an historical right to Kuwait and its oilfields. The invasion was met by almost universal international condemnation and in January 1991 a coalition led by the United States with substantial British and French involvement launched strikes against the Iraqis which resulted in liberation of Kuwait. Saddam's forces were driven back into Iraq by the Coalition and faced annihilation from air and ground attacks on the road to Basra. However, a ceasefire was called at the end of February and most of the Iraqi Army remained intact. Saddam was permitted to remain in power, which in the light of subsequent events was undoubtedly a mistake by President Bush and Saddam turned his attention to suppressing internal opposition to his rule. In the north the opposition was concentrated in Iraqi Kurdistan where the Kurds achieved some degree of autonomy after fighting off the Iraqi Army in the mountainous country above the Tigris-Euphrates plains. Limited aid was provided to them by Britain and America along with other counties of the West, mostly in the form of food supplies and fuel. The towns in the mountains remained largely in Kurdish hands defended by the Kurdish freedom fighters known as the *peshmerga*, but Saddam's military units operated throughout the area as well.

About a week before Christmas in 1992, I received a telephone call from a Ms Joan Anderson who was Gulf Coordinator for Save the Children Fund. She said SCF was looking for help from a British engineer for a structural assessment of a bridge over the river Zab on a relief supply route in Iraqi Kurdistan. For a relief project like this, SCF as an NGO were being supported by the British government Overseas Development Agency (later the Department for International Development and recently merged with Foreign Office) and my name had been mentioned by ODA. She explained there were two routes from Turkey to the south and east of Iraqi Kurdistan, one through Barsan in the east near the Iranian border, and one in the west across the river Zab. The Barsan route was often blocked by snow in winter while the western route was usually clear. The bridge over the Zab on the western route was therefore a vital link for supply of goods. It had recently been constructed replacing an unreliable ferry, but there was some doubt as to whether it was safe for use by heavily loaded trucks. If the route could be fully opened-up through the winter, it would alleviate a lot of hardship.

Although I had little knowledge of Iraqi Kurdistan at the time, I said I would take it on. As there seemed to be some element of personal risk, I did not think I could very well ask Alf Treserden, or a younger, less experienced engineer to go. As we were still in the throes of the financial difficulties, it also offered me a welcome distraction out of the office. Joan said SCF would plan for me to make the visit in early January and as she would be in Scotland over the holidays, she would like to visit me at home to brief me herself between Christmas and New Year. Margo and I sat with her in the lounge while she described the Kurdish resistance to Saddam Hussein and what she knew of the bridge. It appeared a Kurdish commander, (in effect a local warlord) had "liberated" Bailey bridge units from the Iraqi Army and a Kurdish contractor had used the units to construct a bridge over the Zab to complete the supply route. However, air photos taken by the Americans appeared to show it in a state of partial collapse such that it would be unsafe for heavy trucks. An inspection was needed by a bridge engineer to determine the true condition and in view of the shortage of food and fuel in Iraqi Kurdistan this had to be carried out as soon as possible as people in the south and east were close to starvation. I would be accompanied by a British road engineer who had some knowledge of the route and by the local SCF aid programme manager. There was a possibility the Americans might provide a helicopter to take us from the Turkish border direct to the bridge site, otherwise we would have to go by road through Iraqi Kurdistan. Joan advised I had an injection to protect against hepatitis before going. Just what I did not want over the holiday I thought at the time, especially as when I had it for Nigeria it was administered through a large needle in the backside. Finally, she asked about my fee. Given that the visit was to be for humanitarian purposes I said there would be no charge for my time. However, she said there needed to be a nominal fee at least in order to qualify me for £300,000 insurance cover provided by SCF for contractors and SCF would also be responsible for all expenses in making the visit. We settled on a fee of £100 and I confirmed this in a fax. Then, almost as an afterthought she said Princess Anne was a patron of SCF and she might well want to sit in on a debrief when I got back to London.

Arrangements were finalised and in early January I flew to Heathrow where I met the road engineer, Thom Fraser. Together we flew on to Istanbul, stayed overnight, then went on again to Diyarbakir in eastern Turkey. We nearly did not make it beyond Diyarbakir. There had been heavy snow in Istanbul in the morning and our flight was delayed by four hours. When the aircraft finally reached Diyarbakir there was more snow and thick mist. The pilot attempted a landing and just as the aircraft was about to touch down, he slammed on full power, banked steeply with a wingtip almost scraping the ground and pulled away. We went around again and eventually landed safely. It seemed on the first attempt, in the poor visibility, the pilot had been attempting to land on a road outside the airport perimeter fence when he realised his error at the last moment.

We stayed the night in Diyarbakir in a hotel called the Caravanserai which dated back hundreds of years to when it was a stopping point for the caravans on the Silk Road. You had to stoop down to go through the doors into the building as the openings were only four feet high—they were designed to keep camels out. Once one got in, it was, in fact, a very comfortable place. The next morning, we set off by taxi for Zakho on the Turkish-Iraq border, over 200 miles away, and got there in the early afternoon. There were long queues at the border checkpoint, but with the help of an agent of SCF we got through quite easily. First, we went to a United

States mission compound inside Kurdistan about a mile from Zakho where we met a colonel in the US Army. Bad news—no helicopter. There had been some incident in southern Iraq and the Coalition had decided to suspend all allied air movements for a week. We would have to go by road. Before we left, the colonel showed us an oblique air photo of the bridge. In the photograph the structure appeared to be twisted to one side and I could see the reason for the concern.

We went on south to Dohuk where SCF rented some houses, and I was given a room in one of them. Dohuk was on the southern limit of the "safe havens" in otherwise hostile territory. These had been proposed initially by the UK prime minister John Major and had been set up by the Coalition. Now that we had to go by road, I was keen to press on further south the next day, but there was another setback before we could get off. A convoy of aid workers in the east, near Sulaymaniyah, had been ambushed by Saddam agents and one of the aid workers had been killed. Whether this was an isolated incident, or part of a general targeting of Westerners was unclear. SCF were reluctant to move south until the situation settled down. All we could do was wait and listen to bulletins on the BBC World Service for any relevant news. Fortunately, there were no further incidents, and we made the trip four days later.

We were in three 4-wheel drive Mercedes vehicles which had been donated by Germany. I was in one with Thom Fraser. Steve Wilson, the SCF aid programme manager and a British nurse employed by SCF who was to go on to Sulaymaniyah, were in another. The third vehicle contained Sarwat, a local engineer who worked for SCF and two *peshmerga*. This vehicle led the convoy. I was in the back of the last vehicle flanked by two more *peshmerga*, both armed with AK-47 assault rifles. It was drizzling when we left Dohuk and as the road climbed up to a plateau the rain turned to a thin sleet and then light snow. About 100km south we stopped for some sandwiches and to relieve ourselves (the men anyway) taking care not to stray off the road in case of anti-personnel mines in the verges (a favourite place apparently to catch out people peeing). Then we were off again, climbing over ridges where the road was snow covered and down into small valleys which were just damp. We were passing along a featureless plain, for all the world like a Scottish moor, when some low buildings, possibly animal shelters, appeared ahead. Suddenly there was a rapid, stuttering rattle from in front and the Kurdish drivers immediately slewed the vehicles off the road. We were in an ambush under fire. The fire continued, sounding rather like strings of Chinese crackers being set off, only with a flatter sound.

The shots seemed to be coming from beside the buildings, but thanks to the quick thinking of the drivers none of the vehicles was hit and we had some cover from a bank beside the road. At the first shots the *peshmerga* had jumped out of the vehicles and were now returning fire from behind the bank. I had utmost admiration for them and still have. They carried out a classic British Army attack on the enemy position exactly as I had been taught in the OTC over 30 years before. Four of them continued to return fire, while two skirted off to the right in dead ground, reappearing about 100 yards from the ambushers where they could enfilade them. Meanwhile the four moved forward in bounds to about 30 yards from the enemy. At a signal from the leader the enfilading fire intensified and the four charged forward firing from the hip. There had been at least three of Saddam's men in the ambush; two were killed, one surrendered and was taken prisoner and others may have got away when they realised that they were up against determined *peshmerga*. The action was over.

We drove up to the buildings where the dead were lying to one side and the *peshmerga* were questioning the prisoner using an interesting technique which involved firing at the ground close to his feet so that he had to dance up and down to avoid the bullets. I think I might have seen John Wayne doing something similar in a film. The ambushers had a truck parked behind the buildings which was commandeered, and the prisoner was pushed into it before we drove off in our vehicles heading for the bridge. I had only one escort beside me now as the other was guarding the prisoner in the truck. My remaining escort was naturally happy at the outcome and had spare AK-47s taken from the ambushers, one of which he handed to me. I examined it carefully, concerned there would be a round in the breech, but my escort who spoke some English showed me the fire selector on the side of the rifle which was set to safe. I was familiar with the old Mark 4 Lee Enfield rifle from my time in the school corps and stalking in Perthshire. I had also fired the Bren light machine gun on the army range in Perthshire, but I had never seen the Kalashnikov at close quarters. I detached the curved magazine, prised out one of the 7.62mm rounds it held, and slipped the round into my pocket as a memento of the action.

We only had another 20km to go to the bridge. Soon we came to a long queue of trucks on the north-west bank of the river and saw the bridge itself where the river came out of a gorge. The river was running full to the base of the bridge supports. Contrary to the photo in the colonel's office, the bridge looked quite sound with no distortion I could see. There were two spans: a double-single side span over a dip on our side, leading to a larger triple-double main span across the river itself. There were about twenty *peshmerga* and other freedom fighters as SCF liked to refer to them, ringed around in defensive positions guarding the bridge against attack by Saddam's forces. I got out and heard someone say in English, "the engineer", and felt that made the trip worthwhile.

I went forward to meet the contractor's agent who was waiting by the side span. I shook hands with him and realised I was still carrying the AK-47. It did not seem to worry him that I was armed with an assault rifle, however, I said, "I had better get someone to take this so that I can look at the bridge properly". He called to a young lad of about seventeen who was armed with an RPG launcher, but with no gun. The boy was clearly delighted to receive mine to add to his weaponry and I did not get it back. I went over the superstructure of the bridge with the agent and checked the levels across the deck for any twist; there was none. Then I clambered down the steep banks to view the underside of the deck and the parts of the foundations that I could see. The Bailey bridge steelwork appeared to be in virtually new condition and had been properly erected. The diagonal tie bars between the chords were slightly loose and there were only two transoms in each panel instead of four as recommended which would result in some deflection of the stringers under vehicle loading, but these were the only defects I picked up. There were 16 panels of triple-double Bailey Bridge construction in the main span. The first adjective "triple" refers to the number of trusses grouped together on each side of the deck: single would be just one, double is two and triple is three as on Qandil Bridge. The second adjective "double" is the number of standard truss units in the span built up one on top of the other. In the triple-double main span of Qandil Bridge there were two sets of trusses one on top of the other on each side, giving a total of six truss units in all. In Bailey bridges the longer the span or the heavier the loading, the more you use. You do not need to carry out any structural calculations to determine the right configuration as you just use standard tables of span and loading and that is what the agent had done. I had the tables with me, and I checked there and then to find he had got it correct for the main 40m span of the bridge and its single carriageway. With the triple-double configuration on a 40m span, the bridge would be good for a loading of 70 tonnes, roughly equivalent to two heavily loaded trucks.

The agent told me that he had started out to construct the main span only with no side span and with backfill right up to an abutment on the north side of the river. Then there had been a lot of heavy rain and the downstream wingwall of the abutment had collapsed. This would have been caused by poorly compacted backfill with no proper drainage which would have become saturated allowing water pressure to build up causing the lightly reinforced wingwall to fail. There was no sign of distress in the main part of the abutment which was now acting as a bridge pier. This suggested it was stable and safe for long-term use, although I verified the stability by my own calculations from as-built drawings the agent gave me once I got back to Dohuk. After the wingwall collapsed the backfill behind the abutment started to wash out and access to the bridge deck was lost. Somehow when word of this filtered back to London things appeared more serious than was the case and the distorted American air photo added to concerns. Fortunately, there had been a lot of bridging units left over from the haul liberated from Saddam and as the agent was resourceful and a good engineer, he erected these over the gap in a double-single configuration. This side span had just been just completed when I arrived and as a double-single would be good for 55 tonnes on its 30m span.

The inspection took about two hours, and it was after 3pm when we finished. To give everyone confidence I said we should carry out a load test and the agent was game. We got two of the biggest lorries available and loaded them up with soil till they could hold no more—about 40 tonnes in each we reckoned. Then he and I walked out to the centre of the main span and signalled to the drivers to come over. The trucks moved slowly on to the bridge deck and crawled past us to the other side. We smiled at each other and walked back to the north side. In view of the lower capacity on the side span I said to stick to just one relief truck on the bridge at a time and as the first truck in the queue moved forward there was a ragged cheer from the guards. Over the waiting trucks went, one by one, clearing the queue in about half an hour. Apart from telling the agent to tighten up the ties there was nothing more for him to do. I congratulated him on an excellent job. We shook hands again and that was that.

We had to take the nurse to Arbil where she would stay in a SCF house before going on Sulaymaniyah. Our convoy got to Arbil about 5pm leaving the nurse with the SCF manager for that part of Kurdistan. Arbil was much busier than Dohuk and is now the capital of Iraqi Kurdistan. We stopped briefly to buy some flatbread "sandwiches" filled with fried spiced lamb (or goat) along with cartons of tea which we consumed in the vehicles and then set off back to Dohuk. We were anxious to get as near to Dohuk as possible before it got dark as we were worried about another ambush. However, the journey back was uneventful, and we were in Dohuk by 8pm. The cooks rustled up food and we drank some beer which was very welcome after that long and eventful day.

I now shared a room with Sarwat who had a deep and abiding hatred of Saddam Hussein and the Iraqis of the plains. During persecution of the Kurds by the Bagdad government in January 1988, chemical weapons were used in an attack on Halabja and over 6,000 civilians were killed. Sarwat's wife and baby daughter were among those who died. He slept with an old British army issue Webley revolver under his pillow, which he kept, not for defensive purposes he said, but to shoot himself if he were ever likely to be taken by government forces.

The next day I wrote up a preliminary assessment of the bridge for SCF and completed the check on the stability of the abutment. That evening we had a party attended by all who had made the trip south as well as the landlord of the houses SCF were renting in Dohuk. Thom Fraser and I had brought two litre bottles of Glenmorangie with us and there was also plenty beer. The landlord provided the food which included a magnificent fish which he said had come straight out of the Tigris that day and was a centrepiece on the table. He was said to be something of a rogue, but he was a likeable one, with several strings to his bow. One of these was acting as football pools collector in Dohuk which involved gathering money from the punters which he took to a central collection office in Bagdad once a week. The football pools in Iraq were based on the same format of team selection from the top British leagues as used in the UK. At the party he brought in a bolt of West of England cloth. He said he had a lot more in Bagdad where it was in high demand in the upper echelons of the Iraqi government. If I were interested, he could organise a fitting and have a suit made for me. I did not take him up on the offer as I expected I would not be very welcome in Bagdad, but we all got along famously that evening. Only later when I thought about it, I wondered if our landlord had a foot in both camps and might have been involved in setting up the ambush.

Thom Fraser and I were driven to Zakho the next day where a taxi had been arranged to take us on to Diyarbakir. It was a slow journey as there was a lot of snow on the road in Turkey and we were stopped repeatedly at police check points and made to get out of the taxi. It seemed there had just been some incident between the PKK and Turkish forces in the violent guerrilla war which had simmered in south-east Turkey for many years. Each time we were stopped, the driver, who was a Kurd, was frisked roughly by the police, although we were left alone when we held up our dark blue British passports (that was in the days before we were issued with the Burgundy coloured EU style passport). After the third checkpoint, however, I suddenly remembered the 7.62mm round in my pocket and reluctantly wound down the window and threw it out into the snow, not sure what the police would have made of it if they had frisked me. Back in the Caravanserai we enjoyed a good dinner and a few beers watching a belly dancer gyrating around the tables. Then in the morning it was on to Istanbul by air, flying over the deep snows of the Anatolian Plateau followed by a flight to Heathrow. I stayed the night somewhere in central London and the next morning presented myself, still in my cold weather clothes and boots, for my debrief at SCF's office in Vauxhall and waited in the reception area.

After a few minutes Joan appeared with someone she introduced as a director of SCF. I wondered if the Princess Royal might be waiting for us upstairs, or perhaps we would go around to the Palace, but when Joan led us into a conference room there was nobody else there. "Is Princess Anne not here?", I said. "Well, you took a bit longer over there than we expected", Joan replied, "and I'm afraid Princess Anne had other engagements this week. We tried to get Lynda Chalker instead (Minister for Overseas Development at the time), but she is making a statement in the House today. However, they are both very pleased with the outcome. Would you like coffee and biscuits?"

I handed her my notes which she copied and my debrief was over in half an hour. Back in the office about a week later I wrote up a proper report which I sent on to her with the invoice. Job done, but I never did get to meet Princess Anne.

Sullom Voe

Obtaining an appointment as consulting engineers for the Sullom Voe terminal jetties in 1974 had been a great triumph for Peter Fraenkel's new partnership so soon after starting up. The design of the jetties was excellent and very buildable which is so important in marine works. The construction was carried out by a consortium of British contractors and their workmanship was of a high order. All in all, the jetties were a good advertisement for British marine civil engineering at the time, although standards seem to have slipped a bit forty-five years later and there are few British marine civil contractors nowadays in our risk-adverse world.

When I went to see Capt. Sutherland in 1995 to try to get SIC and Sullom Voe back as clients, my only previous involvement there had been designing steel frames to support fenders on the head of Jetty 1 to cater for short gas carriers which had difficulty berthing against the fenders on the dolphins. However, when we were given the job of carrying out a condition assessment of the jetties, I spent some weeks on the jetties myself along with two of our graduate engineers and became familiar with the various facilities. The basic function for which the terminal was designed was to export crude oil from subsea pipelines running from the Brent and Ninian oil fields in the North Sea. There are four jetties complete with berthing and mooring dolphins. Each jetty is accessed from the shore by an approach bridge designed to carry maintenance vehicles and product lines to the jetty head. Jetty 1 had been designed originally as a gas jetty to export LPG while the other three jetties were oil export jetties designed for 350,000dwt tankers. Water depth at the jetties is between 17 and 20 metres. All the jetties and approach bridges are supported on steel tubular piles driven through thin deposits of gravelly sand to hard gneiss rock which formed from granite through a process of metamorphism. Other facilities include offices and an administration building along with radar and leading light towers. Not long after we started again as consultants for the terminal in 1995, Jetty

3 was converted to receive imported oil delivered by shuttle tanker from an FPSO serving the Schiehallion oil field to the west of Shetland.

As described earlier in the memoir, the terminal is owned by SIC and until 2016 it was operated by BP who were one of 26 oil industry partners using the facility. There were annual negotiations between SIC and BP to agree the maintenance expenditure every year as the cost had to be paid for by the oil companies under the terms of the agreement for operation of the facility. The negotiations which I sometimes sat in on were often quite fraught as BP wanted to spend as little as possible, while SIC were intent on keeping their assets in top condition. Maintenance was carried out over the summer each year when the weather was at its most benign (which is not saying much in Shetland) in 3-year term contracts. With Robin Theobald supervising the work things generally ran smoothly except in the year I was in Malaysia when a contractor (recommended by BP) turned out to be useless and as Engineer I had to advise SIC from afar to re-enter the contract. After that we were very careful to select a competent contractor and a local firm called Malakoff, with home advantage, generally saw off other bidders and usually performed well. This maintenance work provided us with good "bread and butter" income each year.

Although the jetties had been constructed to the highest standards, by 2012 some of the concrete in the berthing and mooring dolphins was showing its age. Spalling of concrete was occurring on the undersides of these structures which were exposed to waves and salt spray, and we became concerned that the embedded steel reinforcement from which the structures derive their strength could be corroding. In a marine situation, over time chloride salts from the spray penetrate the concrete and when the salts reach the reinforcement the natural alkalinity of the concrete is lost, and the steel begins to corrode.

Techniques have been developed to determine the amount of corrosion which is occurring without the need for large-scale break out of the concrete to reveal the steel reinforcement, at least initially, and we used these in a comprehensive survey of all the dolphin structures. This was carried out by so-called half-cells surveys in which an electrode forms one half of the cell and the reinforcing steel in the concrete the other. The greater the electrical potential measured by a voltmeter in the circuit connecting the reinforcing bar to the half-cell, the higher the risk that corrosion is taking place. Using this method, we were able to map the areas at highest risk and organise repairs accordingly. This generally involved breaking out concrete on the underside of a dolphin, replacing damaged reinforcing bars and then spraying concrete over the reinforcement to restore the concrete cover to the bars. The work was done by a specialist concrete repair firm over several summer seasons, working as nominated subcontractor to the regular maintenance contractor. It was very successful, and the full integrity of the dolphins was restored.

During the surveys it was discovered that the front skirts on the berthing dolphins were also severely corroded. These support the large rubber fenders against which the tankers berth. There were four berthing dolphins on each jetty and on all of them the skirts were damaged. Obviously, the skirts could not be repaired while vessels were using the dolphins. However, by 2012 the throughput of the terminal had reduced to well below its peak in the 1990s and only two jetties were in regular use. In discussion with BP, it was therefore agreed we could repair all four dolphin skirts on one jetty at a time, while the tankers were diverted to the other jetties, provided the repair work could be completed over a single summer season in each case. We had to find a way to do this which was not easy given the number of operations which would be required. It was also apparent that the large cylindrical rubber fenders and their heavy steel support bars which were hung off the skirts had reached the end of their useful life and should be replaced with more modern fenders at the same time.

The first of the skirts to be tackled were those on Jetty 2. As it would have been extremely difficult and expensive in the use of temporary works to cast new concrete skirts on exposed dolphins located in deep water, I had the idea that we could use prefabricated steel skirts filled with concrete, having first cut away the existing concrete ones with thermic lances. The steel skirts would be supported by raking struts which would extend from the base of the skirt back to the underside of the dolphin to which they would be clamped by high strength "Macalloy" bars. These struts would carry a share of the force generated by compression of the fenders as the tankers were berthing, the rest of the force being taken by undamaged concrete near the top of the dolphin. During the detailed design process, it was decided to modify the scheme and not remove all the damaged skirts, but to take out slots through which the struts could be passed before clamping them to the undersides of the dolphins.

The steel skirt assemblies were made in China at significantly lower cost than comparable fabrication in the UK or Europe, even allowing for shipping. I made several visits to the fabrication facilities which were in a place called Baoding around 70km south of Beijing. The standard of workmanship of the steelwork was first rate and the large fabrication shop in which the skirts were made was well set up with modern cutting and welding equipment. On my first visit to Baoding, I was accompanied by Henry Liu of Mannings in Hong Kong, by arrangement with my friend Mark Cheung, the owner of Mannings Consultants Asia. Henry and I flew up from Hong Kong and we spent two days in Beijing before going on to Baoding. The steelwork fabricator provided a car and a driver, and we were taken around the sights of Beijing with Henry explaining the background to each. On a subsequent visit I was joined by Andrew Insker, SIC's Engineering Manager for SVT, with whom we had an excellent relationship.

Once the steel skirts were filled with concrete and complete with struts, the assembly weighed around 50 tonnes. They were slung into position by a large shear leg lifting vessel hired from a Norwegian company called Eide Lift based in Stavanger, which made the sea crossing to Sullom Voe in weather windows on each occasion it was used. Getting the struts through the slots in the existing skirts and connecting them to high strength bolts passing through the dolphin pile caps made for a lot of difficulties on site, particularly as the piles supporting the dolphins were a little out of tolerance in their positions and got in the way of the struts. As is often the case, what appeared simple on paper was anything but. To get around this, before the shear leg vessel set out, it was necessary to carry out laser surveys of the undersides of the four dolphins to determine the precise position and orientation of the piles in three dimensions, so that fit up went smoothly when the vessel arrived. In retrospect the slot idea gummed things up at a critical stage in the operation and it would have been better to have removed the skirts entirely as originally envisaged for ease of access to the underside of the dolphins. Nevertheless, even with the difficulties, the work was completed within BP's deadline.

Mindful of these problems with struts we made further design modifications for the dolphin repairs on Jetty 3. This time the steel skirts were strengthened so that they could cantilever from fixings in sound concrete at the top of the dolphins without the need for struts at their base. They were fixed to this sound concrete with groups of high tensile bolts which were anchored in the concrete and stressed to 50% of their tensile capacity to provide a clamping action. This made life a lot simpler, and the work was completed with time to spare.

Quotations were obtained from three European fender suppliers for the new rubber cone fenders we had specified. After assessing the technical proposals and financial quotations received, we awarded the contract to a company called Quay Quip who were based in England and the Netherlands. One of the advantages of cone fenders is low reaction forces when the cones are compressed during berthing of vessels. When properly sized to suit the berthing energy to be absorbed by the fenders, the forces also remain virtually constant throughout the berthing process, only increasing as the design capacity of the fender units is reached. The fenders have facing panels made from low-friction material so that lateral sliding forces from contact with the sides of vessels are minimised. The cones and facing panel assemblies are restrained by chains anchored in the quay structure, or in our case in the steel skirts.

Quay Quip obtained their large rubber cone fenders from a factory in Qingdao in China and following one of my visits to the steel fabricator in Baoding, I flew there to witness our cones being tested. These were the largest rubber cones produced by the manufacturer; 2000mm diameter for the outer dolphins and 1800mm for the inner dolphins. The load testing procedure involved applying a gradually increasing load to the cone in a large press and measuring the compression movement of the cone under this loading for comparison with the theoretical movement. The loading was increased up to the rated load for the cone at which it was expected the movement would be within 20mm of the specified value, then the load was further increased until the cone was squashed flat. The loading was then reduced in increments until the cone was completely uncompressed and any residual deflection was noted. All our cones passed these tests within allowable tolerances. I enjoyed the visit to Qingdao, which is a pleasant city on the coast of the Yellow Sea opposite the Korean Peninsula and was shown around the city by the granddaughter of the factory founder, a young lady with the English name of Grace. In one respect Qingdao is a bit like Orkney in Scotland, where if you ask for a whisky in a pub, you are told you can have any whisky you want so long as it is a variety of Highland Park. In Qingdao when you ask for a beer you are told you can have any beer you want so long as it is a variety of Tsingtao. I was not complaining of course, as having played for the Tsingtao squash team in Hong Kong in the 1980s, I had an almost proprietorial interest in the beer.

There were still repairs to be done on the berthing dolphins as their undersides were also damaged, although rather less so than in the case of the mooring dolphins. However, they were at least secured for the meantime and the substitution of softer rubber cone fenders meant the loading on them from berthing of tankers had been reduced.

Smith Quay in Peterhead Harbour

In my early years in civil engineering, I liked to think I could produce original design ideas and innovations which combined clever engineering with economy in construction cost. But I gradually came to realise that truly innovative designs are rare and often some seemingly brilliant idea contains a fatal flaw which renders it unsuitable for the purpose of the project and only becomes apparent as the design progresses. Better just to aim for improvements to tried and trusted design concepts and small enhancements such as the double hinge detail I developed for the naval dockyard gates in Bangkok. However, very occasionally, a genuine cost saving innovation does present itself and can be pursued with benefits all round. The innovation in the design of the Smith Quay in Peterhead was one which not only saved millions of pounds, but also rescued the project and brought large financial reward to the client, and the community in Peterhead.

After re-establishing ourselves with PBA on the design of repairs to the Princess Royal Jetty, Bert Flett commissioned us to carry out a major study into wave disturbance in the harbour and the vessel downtime this caused at various berths and to make recommendations for mitigating measures. The cause of the wave disturbance was obvious—the 200 metres wide entrance to the harbour through the outer breakwater arms admits a lot of wave energy. It was how this energy manifested itself at the berths that needed to be accurately determined in order to decide on possible solutions to the problems created. Guy Olliver, who had come over to us from BMT played a big part in this work. He was ideally suited to it as he had run a major hydraulic laboratory for Wimpey, in the days when Wimpey was a civil engineering contractor of some note, and Guy had been responsible for testing many physical models of harbours throughout the world. Physical models accurately represent actual conditions in the real world so long as they are true to scale in all respects. Running so many physical models had given him an almost unique insight into actual wave conditions in harbours and how adverse waves could be mitigated effectively by suitable works such as breakwaters and revetments. As part of the study a large physical model of Peterhead Harbour was constructed at 1:100 scale and tested by HR Wallingford to a modelling specification drawn up by Guy. The initial purpose of the model was to investigate various configurations of an extension to the outer breakwaters which would reduce wave activity in the harbour sufficiently to cure the problems at the berths.

Unfortunately, the likely cost of an outer breakwater extension which would do this proved to be well above what the Authority could afford. Significant wave heights at the entrance on a 100-year return period were calculated to be of the order of 11m and this would require massive armouring for a rubble mound, or very large concrete or steel caisson units for a vertical wall. However, the information obtained from the model proved invaluable. Using it we were able to investigate more modest improvements through works in specific areas within the harbour itself. One of these was at the so-called Smith Embankment, close to the entrance to the inner harbour used by fishing vessels where new berths might be constructed, although the water depth in this location would have to be increased substantially by dredging if larger vessels were to be catered for. PBA would not previously have been able to develop this area near the fishing harbour as it would have been outside their jurisdiction, but in 2006 PBA had taken over the near bankrupt Harbour Trust. The harbour was now under a single authority and operated as Peterhead Port Authority or PPA. By this time also, Bert Flett had retired, and a new director called John Wallace had been appointed. I became very friendly with John who was an excellent client for us up to my own retirement.

What PPA wanted at the Smith Embankment was to provide an all-weather berthing facility for deep-sea fishing vessels which were too large to be accommodated in the inner harbour, and, in addition, berths for offshore oil industry supply vessels up to 160m in length. Reclamation behind the quay would also provide a working and storage area for the port. We were instructed to prepare a design for such a scheme in September 2007 with a view to putting it out to tender in the spring of 2008. First, it was necessary to determine a suitable layout of the new facility from hydraulic considerations in the harbour before undertaking the civil and structural engineering design. There was an existing short breakwater inside the harbour at the Albert Quay, but numerical modelling organised by Guy, again at HR Wallingford, showed that this breakwater would require to be extended to increase wave protection at the new quay. The modelling also showed the alignment of the extension would be a critical factor as waves reflected from it could cause adverse conditions at other berths in the harbour. After a series of iterations, an optimum alignment of the extension was determined and the design of the rock armouring on the breakwater could proceed. With the chosen alignment waves passing through the harbour entrance would be oblique to the line of the extension and would then go on to dissipate on a spending beach. We recognised that, apart from ensuring other berths would not be downgraded, the oblique wave attack was a good thing for the design of the extension itself; it would reduce the need for the very heavy rock armour up to 25 tonnes in weight which had been used on the existing breakwater at the Albert Quay on to which the waves impinged directly. But by how much the armour size could be reduced could only be quantified by physical model testing. As HR Wallingford had no model basins available at that time, we arranged for the testing to be carried out by the French company, Sogreah, in their hydraulic modelling laboratory in Grenoble using a 1: 50 scale model. Sogreah did a good job on the modelling and were most hospitable on our visits to their facilities in Grenoble where we were treated to long and convivial lunches on each visit. It was shown that a main armour size of 10-15 tonnes with a steep slope of 1: 1.3 would be satisfactory. This is a steeper slope than would normally be used, but it had the advantage of keeping the footprint of the breakwater as small as possible and providing space for vessel manoeuvring. We now had a complete detailed design for the breakwater extension and turned our attention to the design of the quay itself.

 Preliminary designs were prepared based on grids of reinforced concrete beams and in situ concrete slabs supported on steel tubular piles at 5 metres spacing, a typical arrangement for quay decks. This would be an "open" structure with a revetment slope below it which was necessary to minimise reflection of waves at the berth. The revetment slope was designed to be steep like the front slope of the breakwater which was just as well as things turned out. Unlike the breakwater extension, the design of the quay deck was not finalised, but was left to be completed by a successful contractor following a competitive tendering exercise. The scheme was therefore put out to tender as a hybrid contract, where the contractor would become responsible for the design of the quay deck and the Employer, through us as Engineer, would take responsibility for the design of the breakwater extension and other works. This was, and still is, a very common approach for projects of this sort.

With a substantial contractor design element, a three-month tender period was allowed, but unfortunately, the bids received were all above PPA's budget of £27 million. PPA had reserves they could use, they could borrow funds from the Clydesdale Bank, and they had secured an EU grant for £6 million, which gave them £30 million in total. However, their accountant, Stephen Paterson did not want to go above a works price of £27 million to keep some financial leeway. We negotiated with the lowest bidder, a joint venture between a Scottish contractor, RJ McLeod and UK/Dutch company Westminster Dredging, making some changes here and there to cut the cost, but got stuck at a little over £30 million. It looked as if the scheme could not go ahead.

Then I had an idea—for a genuine innovation that saved the day. In the construction of the standard type of quay deck we had put out to tender, which all the bidders had accepted with little modification, there are always difficulties in marrying piling operations with revetment construction. You must first install the piles in stages, which would not be easy at Peterhead with hard rock close to the seabed, then carefully build the revetment around the piles. A particular problem is that the piles must be braced for stability until such time that the desk is constructed to stabilise them, and the bracing makes it difficult to place rocks in the revetment. Invariably, expensive piling equipment such as jack-up barges and heavy cranes for placing revetment rocks get in each other's way and with a lot of downtime involved the plant cannot be used cost-effectively. What if, I thought, we could separate these two operations completely, by eliminating the piles though the revetment and spanning the deck over it to a single row of piles at the front of the quay deck. This would enable marine plant to operate on revetment construction unimpeded by piling plant, which in turn would only need to be on site for the minimum period necessary for installation of the piles once the revetment had been constructed. The quay deck could then be built in a clear span from the top of the revetment to the row of piles. But would it be feasible to span a quay deck over 20 metres when it had to cater for heavy duty loading, as much as three times greater than normal bridge deck loading, for what I would be turning it into was, in effect, a bridge? I did some rough calculations, assuming a composite form of construction of deep steel box girders and concrete slabs, back of envelope really, and it seemed it could be done—just. Then I had an informal meeting with a German engineer in Westminster Dredging I was friendly with to whom I explained my idea. He agreed what I was proposing could be cheaper than the conventional arrangement and said he would get back to me with a rough cost saving on a no commitment basis. I got an answer quickly—at least £3 million saving—which was exactly what I was looking for.

I discussed the idea with Roy Glenton who was sceptical at first, but he bought into it when he had run his own checks. We had an outline design in a week, but a lot more than that would be needed if we were to get a realistic price from the bidders without all sorts of caveats. I then spoke to PPA and explained the idea and the potential cost savings. I said if they could give us a month, we would finalise the new design with a bill of quantities which the joint venture could price as preferred bidder to form the basis of a contract. This was really pushing the boat out and Roy would have to do a great deal of the work on his own as I had the company to manage and could not do much myself other than some checking. PPA went for it, though they asked us to take some financial risk on the time element until the scheme was proven to get the desired saving. Roy put in a monumental effort, and he got a new quay design out well within the month I had asked for.

At the same time, I drafted a revised contract with the breakwater works and dredging unchanged, but with provisions for the new quay design and turned the contractual arrangement on its head to make it a simple employer's design throughout with no contractor design responsibility. There was some risk in this as what we had done would be priced by the bidder for a design which been prepared at great speed with a re-measurable bill which might contain errors that could be exploited. The joint venture repriced the works within two weeks and came back with a price which was just below the magic figure of £27,000. There was still a lot of negotiating to be done, but a contract for the works was finally awarded under a standard ICE contract to the RJ McLeod/Westminster Dredging Joint Venture on 28 January 2009.

We now had to develop a fully detailed design and drawings from which the quay could be built and in doing this keep as closely as possible to Roy's preliminary design which the joint venture had priced, or at least not deviate from it in a way which would cost more money. We also had to keep ahead of the contractor's construction programme to avoid causing any delay which would result in claims. Inevitably some things did require to be beefed up and became more expensive, but we were generally able to make savings elsewhere to balance the increases and with this swings and roundabout approach we had held the price at the quoted figure when the detailed design was completed. The arrangement of the quay was, in fact, quite simple and very modular: a line of tubular steel piles 1.4m in diameter and 10m apart in front of the revetment at the edge of the dredged pocket; a series of steel box girders 1.27m deep spanning 24m from a bank seat abutment at the top of the revetment to the line of piles; and a concrete deck made up of prestressed concrete bridge beams spanning between the steel box girders topped off with in situ concrete. I thought there were some nice touches in the final design such as the rocker bearings connecting the piles to the steel box girders which we designed ourselves and had made for us by an excellent engineering company in Sheffield who precision cast the bearing blocks and the rocker pins. There was also a 40m span catwalk to a separate berthing dolphin at the west end of the quay of which Roy was particularly proud. While the dolphin cost around £1 million, it cut out 40m of deck while maintaining the required berthing length and gave a net saving of £1.2 million.

The contract period was 84 weeks with a start on site in April 2009. Marine works involving dredging and construction of the breakwater extension were completed first in the summer of 2009. In this way protection was provided for the reclamation works over the following winter. During this period, suitable material won from dredging augmented with imported filling was placed in the reclamation area, which was brought up generally to above water level allowing plant to operate on the reclamation without restrictions. The separate operations of revetment construction and pile installation were then carried out over the winter of 2009-2010 and, on completion of piling and the bank seat abutment, the quay deck and adjacent dolphin were constructed. The works were substantially completed on the contractual completion date of 30 September 2010.

There were some highlights in construction of the works which in the main was undertaken most efficiently by the joint venture.

First, there was the dredging which was carried out using Westminster Dredging's backhoe dredger *Manu Pekka*, essentially a powerful hydraulic excavator on a barge equipped with spud poles to provide a stable platform during dredging. The berthing pocket at the quay was 200m long by 30m wide and was dredged to give a depth of 10m. Most of the material to be dredged was rock, varying from fractured, partially decomposed granite at the west end of the pocket to harder more competent rock to the east. The *Manu Pekka* made light work of most of it, but pre-treatment was necessary for removal of the hardest rock. This was done by sub-drilling to 3m below the design dredge level in a suitable pattern of holes and then blasting to fragment the rock for subsequent removal by the dredger. Pre-treatment was subcontracted to a subsidiary of Westminster Dredging with an engaging Finnish engineer on site who made an excellent job of it with the minimum of fuss. The only unfortunate event in this part of the work, which was not the fault of the blasting company, concerned the many seals in the harbour. PPA had insisted we employ an independent environmental specialist to comment on construction procedures for potential environmental damage. One of the specialist's recommendations was in connection with underwater blasting. He advised we should set off a small charge shortly before the main charge for fragmentation of the rock. This was to scare off seals on rocks nearby who would immediately swim away from the area of operations he said, and thus be clear of the blast wave which would travel through the water when the main charge went off. However, seals are curious animals and the opposite happened on the first occasion underwater blasting was carried out. On hearing the small charge and seeing the slight disturbance of the water it caused, some seals slid off the rocks and swam over to investigate. Then the main charge went off and when the water settled down, two dead seals were seen floating on the surface. After that we had a seal watcher on hand when blasting and if any seals strayed into the danger area they were immediately scared off with thunder-flashes.

A second highlight was construction of the large diameter piles in front of the revetment. These were installed in grouted sockets by a specialist marine piling company called Seacore using their jack-up barge *Deep Diver*. With the barge in position at a pile location, a 1600mm diameter steel tube with a sacrificial shoe was lifted into position at the side of the barge through which the pile socket was drilled with a 1570mm bit. The purpose of the shoe was to keep loose rock at the top of the socket from falling into the hole and each shoe was fitted with stops to ensure it would only penetrate the socket by 2m so that it would not intrude into the annulus to be grouted when the pile was in the socket. When each socket had been formed, a pile fitted with temporary end caps was floated alongside then lifted to the vertical and lowered through piling gates into the drilled socket. Once the annulus was grouted the pile was held in the gates for at least 12 hours while the grout set. Finally, the piles were filled with concrete to provide end bearing on the rock and for long-term corrosion resistance. The *Deep Diver* crew achieved a cycle time of just three days per pile and the positioning and verticality of the piles was excellent.

The third highlight was the construction of the quay deck. The steel box girders were fabricated and erected by Rowecord, a Welsh steel fabrication company unfortunately no longer in business, whose workmanship was of a high standard. Once the steel box girders were erected on the pile tops, the modular nature of the steelwork allowed easy placement of the bridge beams, followed by fixing of the reinforcing steel and casting of the concrete in the deck. As a result, the entire deck construction process, from commencement of steel box girder erection to completion of deck concrete, was accomplished in just 16 weeks. Probably the highlight for me in this part of the work was the method we devised of fitting up the rocker bearings on the pile tops and the connection of the box girders to them. The box girders were initially erected on temporary stools welded to the piles and connected in pairs by means of down-stand beams at the front of the girders on which fenders would be mounted. Once each pair of girders were connected like this, they were completely stabilised. After final adjustments to the alignment of the steelwork assembly, the rocker bearings on the pile tops were set to their design level. With the box girders still supported on the stools the upper part of the bearing blocks was drawn up to sole plates under the girders and the upper fixing bolts were inserted, but not tightened. The final lifts of concrete in the piles were cast up to the underside of the bearings and, when the concrete had set, each girder was jacked up in turn so that its weight was taken off the stools. The jacks were then released along with transit clamps on the bearings and as the girders settled down on to the bearings the weight of the girders was transferred through the rocker pins to the pile concrete. Finally, the upper fixing bolts were tightened, and installation was complete.

The works including the quay to the novel design were completed on time and within the budget under an ICE 7th Edition contract in a spirit of cooperation between the parties throughout. These parties comprised just the Employer, PPA, the Contractor, Westminster Dredging RJ McLeod Joint Venture and the Engineer, Peter Fraenkel & Partners Ltd. Apart from the environmental specialist, there were no contract advisers, project managers, third party checkers, quantity surveyors, planning coordinators, Uncle Tom Cobley and all, as the necessary functions were performed by the Employer, Contractor and Engineer in the traditional ICE contract system. Completion on time and within budget of a marine works contract is unusual to say the least nowadays and may owe something to the straightforward, tried and trusted form of ICE contract which was used as well as the good relations and trust maintained between the parties.

Lake Tanganyika Ports

A representative of the African Development Bank approached me late in 2009 and asked if I would be interested in an assignment to advise on the condition of two ports on Lake Tanganyika in East Africa. The AfDB were looking for a port engineer to work as an individual expert for the assignment (or mission as they called it) and I explained I could only take it on if it was done through the firm. This did not suit, apparently, and I expected them to go elsewhere. However, they came back to me a few months later and asked if I were still interested could I give a fee quotation for the mission including field visits and compilation of a report. The field visits would be to Mpulungu in Zambia at the southern end of the lake and Bujumbura in Burundi at the northern end over 400 miles away. This condition assessment of the ports to be financed by the Bank was part of a large transport study into ways to increase the flow of goods from Durban, up through South Africa and Zambia and on to the landlocked countries of Rwanda and Uganda in the north. After some haggling, we agreed a fee and the field trips were set for the end of April 2010.

First, I had to arrange vaccinations as most of the ones I had had for places requiring active prophylaxis were out of date. Scarcely anything was needed for Zambia, but Burundi was another matter and I got typhoid and tetanus boosters, and vaccination for diphtheria, polio, and rabies, the last given over a period of one month in three separate injections in the upper arm. Once jabbed up, I flew to Lusaka in Zambia where I was to stay in the Intercontinental Hotel to start with. It was my first time in East Africa which appeared to be much more ordered than Nigeria. An AfDB representative called Mumina met me in the hotel and introduced me to officials from the Department of Transport, one of whom, Kenneth, was to accompany me to Mpulungu, the Zambian Port.

We set off two days later in a four-wheel drive Nissan, accompanied by the Port Manager, Kusumbe, who split his time between Mpulungu and the Transport Department HQ in Lusaka. Mpulungu is about 1100km north of Lusaka located at the extreme south end of Lake Tanganyika. The journey took us over 12 hours, sharing the driving between us on roads which were well maintained for the most part except for one stretch of gravel road which was in the course of being upgraded. As well as being at the southern end of the lake, Mpulungu is close to the Democratic Republic of Congo, a place you avoid if possible.

I liked Mpulungu as it was full of life, especially the pubs of which there was an abundance and in which Kusumbe introduced me to his friends, Zambian, South African and English, who for one reason or another lived in the town. Kenneth and I stayed in chalets on the shore of the lake (Kusumbe had a bungalow for his use when in Mpulungu) and we ate in a pub each evening which Kusumbe recommended. Zambians like their steaks well done, bordering on the burnt, and on the first evening I commented on this to the cook, a cheery girl who served us as well as cooked the food, suggesting medium rare might be better for me. The next night I said I would have the fresh-water bream from Lake Tanganyika ("English fish" as the locals call it) which was excellent. The cook was disappointed I was not having her steak, but I said I would revert to steak the next night if she could just reduce the cooking time a bit. "Ok", she said, "you show me how it should be done tomorrow, then". So, on the next night I went with her into the kitchen and cooked the steak myself, medium rare, topped with her special garlic sauce, and very good it was, although she shook her head in disbelief to see the pink meat when I cut into it.

The port was west of the town consisting of three jetties which were all very dilapidated and in need of replacement. There were two basic problems which prevented the port working with any semblance of efficiency. First, the water level in the lake had fallen over the years since the jetties were built and it was difficult to access them from vessels at the berths. In Lake Tanganyika evaporation now exceeds replenishment from rain in the catchments surrounding the lake. Run off has been reduced by agricultural development and the hydraulic balance has been further disturbed by water extraction on the Congo side, so that there has been a long-term decline in water level. Secondly, the jetties were far too short for most of the vessels on the lake and there were insufficient mooring bollards on them so that lines had to be fixed to concrete blocks on the shore. Only one of the larger lake vessels could lie alongside a jetty at a time and vessels were double and triple moored in some cases which made them difficult to load and unload effectively. Also, the port could not be accessed by vessels in hours of darkness as there were no leading lights or other navigation aids. With these obvious problems in the port, the report on its condition and what to do about it virtually wrote itself.

One of the vessels in the port while I was there was the *MV Liemba* which had an interesting history. She was built in Germany in 1913 as the *Goetzen*, disassembled and transported in crates by sea to Dar es Salaam and by rail across what was then German East Africa, to Kigoma on the eastern shore of the lake. After this long journey she was reconstructed and launched into the lake from Kigoma in 1915. As the First World War spread to East Africa, she was fitted with 4-inch guns and other armaments to give Germany control of the lake. But in 1916 Britain brought armed launches through the Belgian Congo to attack *Goetzen* and other German vessels in Kigoma. At the same time the Belgians established an air base near the lake from which bombing raids on Kigoma were carried out in combination with the attacks by the British vessels. When Britain cut the railway line from Dar es Salaam and advanced towards the lake by land the Germans abandoned Kigoma and to avoid having *Goetzen* falling into British hands they scuttled her in the lake.

However, the engineers who scuttled her had an eye to eventual salvage and protected the engines with a thick layer of grease. When Britain took over responsibility for the region after the First World War, the vessel was raised from the bed of the lake. Once restored she was renamed *Liemba*, and she went into service on the lake again in 1927. She has operated almost continuously ever since with only occasional refits. Certainly, a testament to German engineering.

In the morning before we left Mpulungu, we went to the fish market where my companions were keen to stock up on fish for their freezers in Lusaka. Mpulungu's fish market is located by the shore of the lake on a sandy beach and boasted a large selection of fish, all freshly caught in the lake the night before. There were piles of anchovies which in Malaysia would have been salted and dried in the sun for *ikan bilis*, and large Nile perch, which I knew as *giwan ruwa* in the Hausa language of Northern Nigeria, translating as "water elephant", an apt description of a fish which can reach 2 metres in length and weigh up to 200kg. There was also the more delicate fresh-water bream, the "English fish" I had eaten in the pub during the stay. The fish which my companions bought now had to be packed in ice for the twelve-hour journey south which involved a visit to the ice factory just outside the town. This all took time, and it was early afternoon before we left which meant a good proportion of the journey would be in darkness. I was not too keen on this, but although it was a long drive and we did not get into Lusaka till 3am, we made it without incident on roads which were largely free of traffic. We stopped once at a transport café for some food and coffee and several other times along the route for necessary comfort breaks. On these, stepping away from the vehicle in the pitch-black African night, the stars were a sight to take one's breath away, with the Milky Way like a bright stripe of white paint across the sky.

After a week back in Lusaka working on my report on Mpulungu Port and meeting Transport Department officials, I was off again to visit Bujumbura in Burundi, the second port in the mission. To get there I had to fly to Harare in Zimbabwe for a connecting flight as there were no suitable flights from Lusaka. Bujumbura Port had been built in the 1950s when Burundi and Rwanda to the north were still Belgian colonies, administered as one country as they had been by Germany before the First World War. It was a much larger port than Mpulungu, with long blockwork wharfs on either side of a harbour basin and two open concrete framed finger piers at the harbour entrance. Despite being nearly sixty years old the wharfs were in reasonable shape which I have found often to be the case with blockwork wharfs provided ground conditions are suitable for that form of construction. The finger piers were barely serviceable with the concrete spalling badly, however, and the one on the west side of the entrance was silted up and only usable on one side. There was only one dockside crane in the whole port which was on fixed rails behind one of the wharfs and as it had a limited travel, most vessels were loaded or unloaded by small mobile cranes.

The Port Authority laid on a small launch for me and we cruised around the wharfs and out past the finger piers to the harbour approaches. As we passed through the harbour entrance into the lake the shoreline receded on either side as if looking out to sea and I could imagine one would soon be out of sight of land altogether. Back in the harbour we were confronted by a female hippo with her calf. She became increasingly angry as we edged past her, giving her as wide a berth as possible as we headed for the landing stage.

Apart from the condition of the finger piers, the main problem in Bujumbura Port was siltation exacerbated by the declining water level in the lake. Although Lake Tanganyika in places is one of the deepest lakes in the world, the northern end is becoming increasingly shallow and there is a pronounced littoral drift from west to east across the harbour entrance. It had been suggested by another engineer that a breakwater or groyne should be constructed out from the shore to the west of the port, and this seemed to me a good idea as it would intercept sediment currently building up in the sheltered water of the harbour. I met the Port Director and his assistant who gave me a list of plant and equipment they would like to obtain to improve the working of the port. This was followed by a meeting with the minister of transport in the government to whom I gave my initial findings which were pretty much in line with the thinking in the government. Again, therefore, the assessment report was easy to write.

Once back in the UK, I quickly completed draft reports and submitted them to the Bank. They were largely accepted enabling me to finalise the reports and invoice for the fee. Things went quiet, but six months later I was asked to go to Lusaka again for a conference between the AfDB and the two nations involved in the transport corridor in Lake Tanganyika, Zambia and Burundi, at which I would give a presentation on how things might be brought forward.

I prepared a power-point presentation and flew back to Zambia with it in November 2011. It was based on recommendations for improvements to the two ports, both short and longer term, which I had made in my assessment reports. I also gave recommendations on how the project might be taken to the next stage which would involve site investigations and surveys. This was broadly accepted, and I was asked to provide a fee quotation for outline schemes based on topographical and bathymetric surveys at the sites. I was keen to get some basic soils information for the schemes and had obtained a price for limited SI from a South African site investigation company with mobile jack-up rigs which could be brought to Mpulungu by road then set up in the lake. Bujumbura was more difficult as I could not contact any site investigation companies there. However, I extrapolated the South African company's quote to allow for transport of their rigs to Bujumbura as deck cargo on a lake vessel plus additional time on site. It was all very broad brush, but at least provided an indicative cost.

In the event I could not get approval for SI work in either port, but both the Zambian and Burundian authorities did agree to carry out topographical and bathymetric surveys to my requirements. It took a long time for the surveys to be completed and results fed through to us, but eventually survey drawings with lake water depths were provided and the schemes were drawn up, all subject to confirmation of soil conditions for foundation design of the recommended quay structures. That is as far as it went. We got paid our fee, but I fear the outline designs have languished in drawers in Lusaka and Bujumbura ever since.

The IWMF

At the end of my career, I became involved in a very large project—in fact, the largest I ever worked on. It came about through my friendship with Mark Cheung of Mannings Consultants Asia Ltd whose project it was and conversations I had with Mark when I was passing through Hong Kong on my way to and from China in connection with the SIC dolphin repairs. Each time I went I would meet Mark for dinner, either at a restaurant of his choice or at the KCC Grill. Sitting in the Royal

Garden Hotel before one of these meeting I thought how nice it would be to work in Hong Kong one last time accompanied by my wife, right at the end of my career, in a place and amongst people we both liked. I mentioned this to Mark and said I could be available to work with Mannings if a suitable project came up and he felt I could help in any way.

I had known Cheung Kwan Tar (Mark) for thirty years since the late 1980s when he worked in our office in Tsim Sha Tsui as a graduate engineer on bridge design. He continued to work in the Fraenkel organisation and its successors after I left Hong Kong until he set up Mannings in 2003. Over the years his business had done very well, and he was now chairman and managing director of the largest local civil engineering consultant in Hong Kong, employing over 150 staff; all achieved by hard work and a combination of entrepreneurial and technical skill. After 2010 when I started to revisit Hong Kong again regularly, he had been helpful to me, providing staff for inspections in China and introducing me to a mainland contractor with whom we got a substantial consultancy job in West Africa. Now I was offering to work for him. It just needed a suitable project.

However, there was one matter with NIRAS to be dealt with, even if something turned up in Hong Kong. By 2016 it was three years since our merger with them. I was no longer MD of the company as I had handed over the reins to Duncan MacKillop whom I recruited as a director after Roy Glenton had left suddenly in a somewhat unnecessary spat with NIRAS. Under pressure from NIRAS, I had agreed that Duncan would succeed me as MD which he had now done. I was finding it difficult come to terms with no longer calling the tune in the company I had run for twenty years, and I had decided I would retire, taking what I could get for my shares in the company.

As soon as I gave my notice, a dispute arose with NIRAS on the value of the shares, which had to be sold to them on giving notice. Under the terms of the sale and purchase agreement I was due a multiplier on the share value so long as I had remained with the company for at least three years since the merger, which I had. The multiplier was applied to the average profits over the three years. NIRAS argued that losses in one of those years reduced the multiplier to zero and all I was due was the equity value of the shares. But there was no mention of losses in the wording of the agreement, only profits, and while I accepted losses had arisen in one of the years, I insisted the average had to be based on the two profitable ones with the loss-making year struck out. Eventually, after some reasonably amicable negotiations, NIRAS accepted my arguments on the multiplier and the matter was settled satisfactorily in my favour. I was pleased with this, as apart from the enhanced payment I received, it would have been a pity to leave on a sour note when the merger had been largely successful as I saw it and of benefit to both sides.

With the NIRAS matter settled I still had to work out the remainder of my notice period and very fortuitously Mark Cheung came up with a suitable project for me and I was able to do this in Hong Kong with Mannings. The project was the design and construction of an incinerator for burning solid waste and included all the civil engineering necessary for formation of the site. Its official name was the Integrated Waste Management Facility or IWMF. Hong Kong had had an incinerator before, located on the old Castle Peak Road, and I can well remember the brownish haze which rose from it and seemed to form an extra cloud layer above the hills of west Kowloon. The old incinerator was long gone, but the pollution which had emanated from it had given incinerators a bad name in Hong Kong and although the new one would employ the latest technology and produce almost zero emissions, nobody wanted it anywhere near them. Accepting this, the government had gazetted a site for the new one well away from the urban areas, off an island named Shek Kwu Chau south of Lantau. Around 14 hectares of reclamation would need to be formed in deep water and on a soft seabed in order to house the incinerator and its ancillary facilities. Directly in the path of typhoon waves, it was undoubtedly one of the worst sites one could find for such a facility in terms of cost and difficulty of construction. But it was remote and that trumped all other considerations. Construction on that site would be a massive civil engineering challenge and I was keen to get involved.

Consortia were formed to bid for the contract and with great skill and diplomacy Mannings managed to stitch together a joint venture between a mainland Chinese incinerator operator, China National Environmental Protection Group (CNEPG) and Paul Y Construction Company Limited, a local civil and building contractor in Hong Kong. Mannings would provide consulting engineering services to this joint venture and carry out the necessary tender design for them. In due course, prequalification documents were issued by the government to the various consortia which had formed and the CNEPG Paul Y JV made it to a short-list of four who were invited to bid for the project. That was when I came in.

My role in the IWMF was to be the design of the marine works. These comprised a reclamation of approximately 14 hectares, a 500 metres long seawall to protect the exposed seaward side of the reclamation from wave attack, a main breakwater also 500 metres in length to restrict wave heights in a harbour for waste transfer vessels and a quay for the vessels in the harbour. There was also a smaller secondary detached breakwater, to complete protection to the harbour. In addition, the marine works included the necessary ground improvement to support the seawall, breakwater and quay structures and geotechnical works to control settlements in the reclamation fill. In this context "design" meant a tender or bid design which was sufficiently detailed to be priced by Paul Y who would be responsible for the civil and building work and with whom I would liaise closely on construction methods.

As I was still an employee of the company when my involvement started, I had agreed a monthly fee for NIRAS Fraenkel which was only a little above cost, but with a 100% success fee on top if the joint venture won the contract. The monthly fee would be paid to the firm up to the date of my retirement after which it would go to me personally, although the success fee would still go to NIRAS. The arrangement reflected the reality of the situation that to get the work the consultants involved had to take some financial risk. In addition to the fee, I received a per diem payment for personal expenses in Hong Kong and Mannings arranged a service apartment close to their office (and, importantly, the KCC). They paid my international air travel and kindly also paid for travel for my wife. All in all, it was a very satisfactory arrangement for us, and I believe it was fair to all the parties involved.

We arrived in the summer of 2016, but issue of the final bid documents for the IWMF was delayed and initially I assisted in vetting health and safety plans prepared by Paul Y for another project, a massive reclamation for a third runway at the international airport at Chek Lap Kok. At 700 hectares it dwarfed the 14 hectares of reclamation for the IWMF, but in other respects it was simpler as the water was relatively shallow and wave exposure at Chek Lap Kok was almost benign in comparison with conditions at Shek Kwu Chau. The airport reclamation was an engineer's design, and one benefit of my involvement was access to the specification for forming the reclamation on soft ground which helped me when I came to the design of the IWMF reclamation.

The IWMF bid documents were finally issued to consortia on the short list by the government in November 2016 with a return date for tenders in April 2017. Unusually, the client was the Environmental Protection Department (EPD) rather than Civil Engineering & Development Department (CEDD) as might have been expected with the high civil engineering content in the project. The EPD had never managed a large project before, but as the incinerator was the key component of the project the EPD was in charge. Their consultant was AECOM who had produced only basic scheme drawings, but a rather detailed specification. I did not concern myself with the contract terms as my involvement was to be purely technical, but I understood these were particularly onerous for the contractor with potential for unlimited liability. On reviewing them, one of the other bidders pulled out, to leave only three of us.

On our side subconsultants had been engaged for specialist roles relating to the incinerator such as electrical and mechanical works and architecture. I recommended that HR Wallingford (HRW) was brought in for assistance with hydraulic aspects and wave assessment which were likely to be key issues in the marine works design. I had worked with HRW many times over the years for projects in the UK and overseas and I knew they could be relied on. I also felt that their inclusion in the team would enhance the credibility of the bid as they were well regarded worldwide, especially if they could be persuaded to commit to our group on an exclusive basis which they did after some lobbying on my part.

Looking back, it seems to me that the influence of the EPD resulted in unnecessary engineering difficulties at an already difficult site with no real environmental benefit from the restraints which were imposed. The artificial island on which the incinerator was to be built had to be located 100 metres off the south coast of Shek Kwu Chau with a clear channel between the natural shoreline and the artificial island. The reason for the channel seemed to be to avoid disturbing coral clusters just offshore, but with proper precautions these could have been moved lock, stock and barrel to a suitable site elsewhere. The artificial island could then have been built out directly from the shoreline which would have avoided the need for a separate retaining structure on it and an unnatural narrow channel which will possibly silt up in time.

The main problem, however, was a restriction that had been placed on dredging, although this was not clearly spelled out in the tender documents and there was some confusion to start with. The seabed was composed of soft marine clay which in places was up to 15 metres thick over more competent alluvium. In our design of the Trunk Road embankment in Tolo Harbour most of the soft material was dredged out and replaced with submersible filling so that settlement would not be an issue. But things had changed since then and now dredging was frowned upon in Hong Kong as damaging to the environment. For the IWMF limited dredging in certain areas was permitted if silt curtains were deployed to contain fine particles within the site area, but the allowable volumes which could be dredged each day were so low that for practical purposes no significant capital dredging could be carried out. Clearly the soft clay here could not support seawalls or breakwaters, no matter what form they took, and without dredging, the only realistic option was improvement of the clay in situ. This would almost invariably mean deep cement mixing (DCM) which had become popular for soft ground improvement throughout the Far East. However, DCM treatment requires large amounts of cement worked into the ground using heavy plant for the DCM operations and I wonder whether the environment might have been better served by allowing dredging. Had the EPD or AECOM carried out a proper comparison of the methods to obtain an energy balance between the two, or was it just accepted that dredging was bad per se? Who knows, but that was the situation and the bidders had to accept it.

A somewhat less critical issue, but nonetheless an annoying one, was the alignment of the seawall and breakwater which had been specified. The tender documents required these structures to be set around a curve to a given radius which could not be changed in the slightest because the plan position of the structures had been gazetted in precise detail. In deep water with year-round wave exposure, it would have been a lot easier to build the marine structures in straight lines. There seemed no obvious hydraulic reason for curving the structures in plan and to effectively fix the alignment before detailed design was carried out made no sense. In this context gazetting was analogous to a UK Parliamentary order for works in new and existing harbours and these are always drafted with wide tolerances in the layouts to allow for design changes. Whether the fixed layout was down to the consultants or the EPD we could not say, but it made the tender design of the structures more difficult than it need have been.

But these obstacles were not insurmountable and by November 2016 we were ready to go. The biggest decision to make as far as the marine works were concerned was the form of the seawall and breakwater. There were two basic options. They could either be vertically sided structures or rubble mounds. I had designed sea walls and breakwaters of both types, but in deep water subject to high waves and swell a rubble mound would require a large base width footprint for stability and with the extensive soft clay, a much wider zone of DCM treatment would be needed than with a vertical structure. This would be likely to render a rubble mound solution uneconomic. For vertical structures there were two alternative forms which are commonly used: large steel ring cell type structures which are sunk side by side through the clay and alluvium to sufficient depth for stability, or concrete caissons cast in a suitable dock and floated to the site before being sunk on prepared bedding on the seabed. Successful vertical seawalls and breakwaters have been constructed as both types, but there had been problems when ring cell construction was used in Hong Kong to retain a large reclamation near the airport. This was for a border post at the Hong Kong end of the Zhuhai-Macau-Hong Kong Bridge and the works there were much delayed while difficulties with sinking the ring cells were overcome. I favoured concrete caissons over steel ring cells as I believed there was more certainty in their design in terms of overall stability when retaining filling and less risk of damage during construction in an exposed site in deep water. We agreed to adopt concrete caisson construction therefore for both the seawall and the breakwaters.

Many years earlier I had been involved in the design of concrete caissons for a 500 metres long quay on Nhava Island in Bombay Harbour, for the Oil & Natural Gas Commission of India on an oil supply base, but in that location only small local waves were experienced; and neither were the water depths or the ground conditions there particularly challenging. At Shek Kwu Chau, however, things were at the other end of the scale in terms of difficulty, and it was a challenge to produce a successful design. The winning bidder was expected to carry out both numerical and physical wave modelling in the detailed design, but it was recognised that there would not be time for this in the four months initially allowed for preparation of tenders and AECOM had provided basic wave characteristics at the site for various return periods along with coincident water levels in typhoon surges. We assumed AECOM had carried out some modelling to obtain the wave data and requested that this be released so that HRW could review it, but the request was ignored. All we could do therefore was to base our tender design on what we were given which was significant waves of 8 metres and a peak period of 12 seconds.

Paul Y intended to have the caissons manufactured at a yard on the Chinese mainland some way up the Pearl River from which they would be brought to the site on a semi-submersible barge which had a weight limit of 3,200 tonnes for the caissons. The seawall caissons had to retain the filling of the artificial island and the critical loading case for them was at wave troughs when net water pressure acted outwards to combine with lateral earth pressure from the fill. Additionally, they had to protect the reclamation from flooding in wave crests and would be surmounted by wave walls for this purpose. The breakwater caissons were free standing and had to be stable on their own although we could allow some overtopping to reduce the peak wave forces on them. I derived wave pressures on the structures from PIANC Working Group 28 Report. Working through structural stability calculations, the seawall needed to be 28 metres wide and the breakwater 36 metres to ensure adequate sliding resistance. The tender design proceeded on this basis.

In the design, most of the caissons for the seawall were to be founded on a bedding mound at -10mPD with some at a higher level in shallower water near Shek Kwu Chau. The top of each precast concrete caisson after installation was at +2.5mPD with a maximum wall height of 12.5 metres. The seawall caissons were then to be made up to a main crest at +7mPD using in situ concrete with wave walls above this. These caissons comprised eight internal cells with inside dimensions of 6 metres x 6.6 metres making up the overall width of 28 metres The breakwater caissons were also to be founded at -10mPD with the top after installation at +2.5mPD and with the same internal cell configuration These would then be made up to +5mPD with in situ concrete. There was no need for wave walls in the breakwater caissons where overtopping could occur, and they were to be left with flat tops except for drainage falls.

The floating characteristics of the caissons for installation were at least as important as their structural stability and I was greatly helped by Martin Young of

HRW who carried out calculations for me on metacentric height and floating stability. We needed to be sure that the caissons could be floated over the foundation mounds with some clearance prior to installation with a reserve of metacentric height over centre of gravity and we had to have adequate freeboard at the open cells while they were in the floating condition.

Martin also provided invaluable advice on installation methods from his previous involvement in a large caisson breakwater at Costa Azul in Mexico. Based on his experience there we planned for the caissons to arrive on site on Paul Y's semi-submersible barge from which each caisson unit would be floated clear of the barge some distance from its final location. Winch cables would then be connected between the floating unit and the units already installed and the floating unit would be winched to its final position in a partially ballasted down condition with a small clearance over the bedding mound. A simple fender in the form of a gravel bag would be placed on the seabed against the last unit installed and the floating unit would be winched in against this fender to absorb its kinetic energy before it was fully ballasted down on to the mound. The cells would then be filled with water followed by sand ballast to stabilise the caisson. We expected that caisson units would only be transported from the casting yard to site when no typhoons or other storms were likely to approach Hong Kong and there was a clear weather window for the installation time plus a safety margin for any potential delays which might arise. However, if conditions were to deteriorate unexpectedly with excessive waves or swell, a caisson could be sunk on the bedding mound some distance clear of the other units on the mound and re-floated for final installation when conditions improved. A de-ballasting system would be provided in each caisson for this eventuality. By February 2017 we had the caisson design completed sufficient for tender purposes and I could turn my attention to ground improvement to carry the loading arising on the foundations from the seawall and breakwater structures.

The ground pressure at the base of the caissons where they sat on the bedding mound would be approximately $300kN/m^2$ and it could reach up to $400kN/m^2$ at the breakwater caissons under maximum wave loading. With these pressures some form of ground improvement was essential as there was up to fifteen metres of soft marine clay over much of the site. When dredging cannot be used for environmental reasons, the preferred method of improvement in Hong Kong had become deep cement mixing and this is what we proposed to use under the seawall, breakwaters and quay. The method involves extraction of the weak soil by auger to a predetermined depth at which a binder, usually cement sometimes mixed with pulverised fuel ash, is injected into the hole to mix with the natural soil and form, from the bottom up, a column of hardened material. These DCM columns, usually around 1.2 metres in diameter are arrayed in overlapping clusters in lines across the zone to be strengthened.

Design of DCM works was something I had not done before, nor had Mannings, and we needed to get an understanding of the principles of the method as quickly as possible to incorporate it in our marine works design. We approached this in two ways: first, to read up on the theoretical background of the use of DCM for improvement of weak soils, and secondly to talk to contractors who were familiar with the method for practical information on DCM installation procedures.

In terms of the theoretical background, I consulted English translations of various Japanese papers on DCM as Japan was where large-scale use of DCM had originated. I also obtained a very useful design manual issued by the Federal Highways Administration in the USA, entitled *Deep Mixing for Embankment and Foundation Support*. This manual was written in plain language and explained clearly how three-dimensional blocks within soil treated by deep soil mixing (as the Americans called it) could be analysed as an engineered material with structural properties which depended on the ratio of treated soil in clusters to the total soil volume in the block. The formulae given in the manual for use in design were empirical, but they could be relied on as they were derived from extensive laboratory and full-scale testing. External forces on a DCM block, such as we would have from caisson loading, were carried through the block to underlying harder soils, by structural strength within the block. I found this American manual invaluable and with my own structural background I was able to relate to the design procedures it described.

At the time when we were preparing the IWMF tender design, ground improvement by DCM was being carried out in a much larger scale for the third runway reclamation. There were a limited number of specialist contractors experienced in large-scale DCM works, mostly Korean or Japanese, and virtually all of them were involved in the airport runway work to a greater or lesser extent. As a result, there was no immediate availability in Hong Kong of the rigs needed for DCM work on other projects. It was hoped that at least some of the airport contracts would be completed by the time the IWMF was ready to start on site, freeing up heavy auger rigs in use at the airport. One of the DCM contractors involved in the airport work was Dong Ah Construction from Korea who had obtained several DCM contracts relating to the runway. Henry Liu who I knew well from his time in Mannings and our trip to Beijing was now employed by Dong Ah and we were able to tap into some their practical experience of the method through Henry. After discussions also involving Paul Y, Dong Ah agreed to provide Paul Y with a quotation for DCM on the IWMF, contingent on some of their airport contracts being finished so that DCM plant would be available. Their quotation was based on an outline design they prepared for the ground loadings under the caissons which I gave them. I was able to check the Dong Ah design against the design criteria in the American manual and subsequently incorporate the key points in a design basis statement for submission with the tender. At the end of this process, I felt our tender design combined theory and practice of DCM very well.

The last major part of the tender design to be dealt with was the reclamation. Again, this was a mixture of the theoretical and the practical. The reclamation behind the seawall was mainly sand fill forming an artificial island on which the incinerator and its associated facilities would be built. Apart from the weight of the reclamation acting on the soft ground it had to be designed to support a theoretical surcharge of 2 tonne per square metre within settlement limits which AECOM had specified.

Marine deposits like those off Shek Kwu Chau form the seabed over large areas of Hong Kong waters and consist of uniform very soft to soft clayey silts. The deposits at the site were between 8 metres and 15 metres thick overlying alluvium. These deposits could not support a reclamation any more than they could support the caissons. As large-scale dredging to remove the soft slay down to the alluvium was not permitted, the properties of the clay would have to be improved in situ sufficient to carry the weight of the reclamation. Of course, the DCM zones under the caissons could have been extended back to cover the entire footprint of the artificial island and would have provided an excellent foundation on which to place the reclamation, but the cost would have been prohibitive. Two cheaper and well understood methods were available, either on their own, or in combination: stone columns extending down to the harder stratum could be formed in the soft clay like DCM columns, but cheaper; alternatively, preformed vertical drains (PVDs), often referred to as wick drains, could be inserted into the weak layer, coupled with surcharging of the surface of the reclamation which would squeeze water out of the soil to be carried up the PVDs to consolidate the soft layer.

I was not keen on stone columns as I had seen them fail at first hand in an area of very soft ground at the Port of Tilbury. Another consultant had designed an improvement scheme for this area using stone columns with disastrous consequences. After installation of the columns, they had burst sideways when a surcharge was applied, and the soft ground area had become virtually unusable for dock operations except for storing imported cars.

The question was what spacing should be adopted for the PVDs. Clearly, the closer together they were the better to reduce drainage paths in the clay, without incurring too much of a cost penalty. The spacing was also dependent on the placing sequence and height of surcharge on the reclamation and the duration for which the surcharge would be left in place. Another factor was the method of placing of the filling at the start to avoid the formation of mud waves in the soft clay as I had seen on two of the fill areas on the Trunk Road project over thirty years before. I sketched an outline scheme for discussion with Paul Y, but detailed calculation of PVD spacing versus gain in shear strength in the clay was beyond me. As Mannings' geotechnical engineers were all involved in other work at the time, they arranged to have Halcrow (by then CH2M Hill) perform detailed calculations for us on the scheme I had prepared. Halcrow did a good job and we ended up with a workable arrangement of PVDs with the closest spacing adjacent to the seawall where the depth of soft clay was greatest. The maximum height of surcharge Halcrow proposed was 8 metres. This would be applied in stages and moved progressively across the reclamation from areas which had achieved the necessary consolidation, as demonstrated by cone penetration tests (CPTs) at depth in the clay, to areas still to be fully consolidated.

With the completion of the reclamation design the three main elements of the marine works, namely, seawall, breakwaters and reclamation were ready for the tender submission. The design of the quay wall was also completed. It had required little design effort as it was to be a standard Hong Kong blockwork wall on a rubble mound over a DCM zone. The last job was to determine the height of the wave wall at the seawall to minimise overtopping in maximum wave conditions. In their hydraulic laboratories in Oxfordshire, HRW set up a preliminary model of the seawall in a wave flume at 1:20 scale and ran a series of 2-D tests with different wave wall configurations. It was found that 4 metres height of wall was needed, taking the elevation at the top of the wave wall to +11mPD. While this would have to be checked again for random waves in a 3-D model in the detailed design stage, it enabled the design to be finalised for tender purposes. HRW had been of great assistance throughout the tender design and were a pleasure to work with.

I derived great satisfaction from working on the IWMF project which was probably the biggest job of my career. That it was in Hong Kong made it even better for me. My wife and I were pleased with the arrangements Mannings had made for us even though our apartment had just basic facilities. It was in Chatham Court, about 5 minutes' walk from the office and another 5 minutes to the KCC. Although I worked from 9am to 6.30pm Monday to Friday and on Saturday mornings, I could often get back to the flat for lunch and Margo would sometimes join me for *Yum Cha* with Mark and his other directors. On Saturday afternoons and Sundays, we could take a ferry for a walk on Lantau or Lamma Island, or a bus out to Sai Kung. We would eat in the KCC two or three times a week. In the first summer and again as the weather warmed up in the spring of 2017, we would swim in the KCC pool. If not going to the KCC, we would often walk down to a pub at the Kowloon-side waterfront with views of Hong Kong Harbour for a quick drink in the early evening before returning to the flat to cook up dinner in the small kitchen. On weekends, we sometimes stayed in a hotel near Tai O at the far west end of Lantau called the Old Police Station. In the early days of British rule in Hong Kong the police had patrolled the Hong Kong waters of the pirate infested Pearl River estuary from a jetty below the police station. Now the police station building had been converted to a "heritage" hotel in a government scheme to preserve historic buildings, and it was a comfortable, if rather pricey place to stay.

Freed from financial concerns in running NIRAS Fraenkel, I was able to concentrate solely on engineering and rediscover old design skills neglected in years of management. It was a very happy time for both my wife and I and a good way to end to my career.

Epilogue

> We shall not cease for exploration
> And the end of all our exploring
> Will be to arrive where we started
> And know the place for the first time.
> *From Four Quartets, T.S. Eliot*

By early May 2017, I had finished the marine works design for the IWMF, and Mannings' draftsmen were preparing drawings to submit with the tender. After requests from the remaining tenderers, the EPD extended the date for return tenders to 12 July. As my part of tender design was finished, I started to assist in editing other sections of the design basis statement to be submitted with the tender and with some difficulty was working my way through text on incinerator operations. This had been prepared by CNEPG in Chinese, then translated into English before it came to me. It was disjointed, repetitious and confusing. CNEPG were no doubt skilled operators of incinerators on the mainland, but they had never bid for projects outside of mainland China until the IWMF. It seemed they had no understanding of the technical assessment process to which their submission would be subject and the need to convince the assessors that they would be capable of designing and operating the incinerator as envisaged by the client in the unforgiving commercial world of Hong Kong.

Then, out of the blue towards the end of May, my wife started to experience persistent abdominal pains which seemed worse at night when lying down. She was booked to fly back to the UK on 31 May, while I was to stay on in Hong Kong until 12 July, when the tenders were to be submitted. After that we might or might not return to Hong Kong for me to work on other projects for Mannings, perhaps on a part-time basis. On 1 June when Margo arrived back in Scotland, she arranged an appointment with the doctor and, unusually, got one for later that day. We had speculated that the pains might have been related to gall stones and the doctor thought this as well, but blood tests indicted a possibility of something more serious and she went into hospital for investigation. An ultrasound scan revealed inoperable pancreatic cancer. I flew back to Scotland as soon as I could arrange a flight. Our last halcyon days had ended. Margo only lasted six more weeks and she died on 15 July. We had been married for just three weeks short of 45 years. Still bereft after this personal tragedy, it was difficult at first to start on this memoir as Margo had been a constant and vital presence through most of the story. But writing it was therapeutic in that it brought back so many memories of our travels, the difficulties we faced and overcame together and the fun we had in doing so.

Writing the memoir also crystalised in my mind certain views on my own career and on civil engineering in general.

The Firm's Collapse

First, as the troubles PFP (or PFI as it had become in the late 1980s) suffered had an adverse effect on my own life, it might be asked what caused things go so far wrong with many negative factors in play at the same time. How could this have come about in a few short years from a peak of success. I have no wish to denigrate the reputation of Peter Fraenkel as I retain great respect for him and what he achieved and *mortuis nihil nisi bonum*. He had tremendous success in the first ten years of the firm's existence and secured many large commissions from a standing start which was no mean feat. But in my view, and I do not think it is speaking ill of him to say it, he just went on too long as the senior partner with virtually all the equity in the business and with it the ultimate control of the organisation. Some of the success of the early days of PFP was not down to him alone and was in part a result of effort and flair in other people in the organisation. To keep those people, a clear line of succession to the top was needed otherwise they would simply look elsewhere, and without one that is what happened. No doubt, while I was in SE Asia and Hong Kong and remote from events in London, there were internal disputes with the partners and senior staff who were driving things forward with him which I was insulated from, and when they left, they were replaced with people of lesser stamp who did not offer the same challenge to his authority. Unfortunately, with that policy there comes unquestioning acquiescence in decisions made by one person, which may not always be the right ones, as the litany of mistakes demonstrates. It is understandable that Peter Fraenkel would have wanted to retain absolute financial control, as he had put up his own money in the first place, but he and his family might have been better served if he had arranged to step back from management while people with flair were still around. The situation which developed at a time when other consultants were doing well could serve as a case study on how to ruin a well-founded organisation in a short space of time. Thirty years later, Bill Gates, the founder of Microsoft, showed how it should be done when he stood down from the Microsoft Board at the age of 65, selling nearly all his shares in the company and retaining just 1% of the equity. Nevertheless, for all that, I grew to like Peter Fraenkel a lot as a person and think we worked well together after we took back control of the marine civil engineering part of the business and the dark days of the late 1980s and early 1990s were behind us.

The worst of the decisions and the one which brought the firm down in the late 1980s, was to focus on UK highways projects obtained with a series of low bids. Highways work had been carried out profitably overseas in Nigeria, Sabah and Hong Kong, but the commissions had been won on technical merit on World Bank or Hong Kong government fee scales which provided adequate remuneration for the consulting engineers involved. The situation in the UK was very different: price competition for professional services was all as the Department of Transport had thrown the old fee scales out of the window. Given the success of the overseas projects, and the abundance of road schemes in the UK, it must have seemed tempting to grab as much of the low-hanging fruit as possible. Unfortunately, the work obtained at the low fees could not sustain the organisation and mounting losses were inevitable.

Things were made worse by the poor calibre of engineers recruited to carry out the work. A cadre of highly experienced road engineers had worked on the overseas jobs. Roger Base, John Smith and Colin Campbell in Nigeria, John Smith again in Sabah, and Tony Bowley in Hong Kong were all excellent road engineers with a wealth of experience. For the UK jobs which were extremely demanding only Colin Campbell remained. He was as competent as the others, but with limited support from inexperienced staff around him he had no chance of success. All in all, it had been a recipe for disaster.

Related to this, there is the question why I did not get out when I came back to the UK and saw the state of play with the company and all the problems besetting it. At first, I could scarcely belief the mess the firm I had got into and once that became clear to me, I could have confronted Peter Fraenkel and told him I had been brought back to the UK under false pretences, and I was returning to Hong Kong or quitting the firm altogether. I may have felt constrained by the six months' notice period I was on, but in retrospect what I could have done was get myself another job and told the firm to whistle for the notice as it would have been unenforceable. However, as a family we had made a lot of arrangements for the return which were not easy to unravel. Perhaps by default, and some misplaced loyalty, I made the decision to remain and muddle through and chances to move on diminished as I grew older myself. Before long, it became impossible, and the only option was to try to build the business up again once I had some measure of control and this was partially successful at least.

Civil Engineering: Its Strengths and Weaknesses

Civil engineering is something I drifted into without much thought or knowledge of what it entailed, but. I enjoyed working as a civil engineer. As my father said of civil engineers, I got out and about a lot, and I travelled over much of the eastern part of the world and left various structures as a mark of my presence on the surface of the planet. I derived a lot of satisfaction in developing schemes for clients and seeing them built to my designs. In some cases, I had to make modifications to the schemes when conditions turned out to be different than expected, which is what I consider engineering to be about. It is an inescapable fact that there is always some uncertainty in the outcome of civil engineering works, and particularly in the final cost. That is not surprising as the physical environment in which we work has a great deal of potential for the unexpected which can frustrate those involved: weather conditions may be unusually bad, ground properties may present geotechnical problems not anticipated and availability of construction materials and their costs may be adversely affected by changes in market conditions outside of our control, to name just a few. Mindful of this, it is important that a client chooses wisely in his engineering consultant and appoints one with real experience of the type of works to be constructed so that such problems can be anticipated and dealt with sympathetically if they do occur. Equally important is to ensure that bidders for the construction contract are also suitably experienced in the work to be carried out. Contractual disputes between employers and contractors will arise if things go wrong, but with an understanding engineer and a responsible contractor it should be possible to resolve disputes in a satisfactory way. It is this flexibility and inventiveness which distinguishes civil engineers from other professions and has contributed so much to society.

However, during my career I became frustrated at some aspects of the way the profession organises its affairs. For many years since the nineteenth century civil engineering contracts in the UK had been written around three parties; Employer, Contractor and Engineer, the latter a named person or firm with substantial powers under the contract. This system was enshrined in the ICE Conditions of Contract which placed the Engineer as a third, independent party in the commercial relationship between Employer and Contractor, an unusual situation in law, but one that had seemed to work until the 1980s when increasing numbers of contractual disputes prompted calls for change. Those seeking to improve the system and proposing remedies for these disputes focused on the role of the Engineer as one cause of the problems, questioning his independence and impartiality and implying that often the Engineer would simply do the Employer's bidding, rightly or wrongly, disadvantaging contractors and aggravating disputes. Out of reports by Latham in 1994 and Egan in 1998 came the New Engineering Contract which did away with the Engineer's role altogether, replacing it with that of Project Manager, Designer and Supervisor. Since the Project Manager was beholden to the Employer under this contract and the Designer and Supervisor had no direct powers, one might think this was throwing out the baby with the bath water to little end. Be that as it may, what I found incredible was the acquiescence of the Council of the ICE in readily endorsing the NEC even as it was doing away with the role of the Engineer, something which must inevitably have reduced the status of civil engineers in the construction industry. If there really were faults in the way we engineers approached the contractual role of Engineer, surely a tighter, more constrained role for the Engineer could have been found to remove any tendency for bias and avoid the need for such a drastic step. I wonder how many of the Council members at that time had acted as Engineer on contracts themselves and had real first-hand experience of administering them properly under the ICE Conditions to the benefit of the works.

It is true that the NEC has become popular, particularly with contractors (one can speculate why), so much so that the ICE itself sold off its rights to the ICE Conditions. These were taken up by the Association of Consulting Engineers, the same toothless organisation which had done little to help its members when their livelihoods were being threatened by the government's abandonment of the UK roads programme in the 1990s. The ICE Conditions were then renamed the Infrastructure Conditions of Contract in which the clauses of the ICE Conditions remained largely unchanged, but the damage had been done and the NEC became the contract of choice throughout the civil engineering industry.

For all the support the NEC now enjoys from government and large corporations, there is no evidence using it has led to a reduction in contractual disputes which seem as common as ever. Neither are there any fewer cost overruns, and in fact it is now rare for a major project to be completed on time and within budget. On both counts the NEC has failed to make any improvements and its weaker controls compared with the ICE Conditions may have made matters worse and allowed more sub-standard work to slip through the net.

Another aspect of contract administration which should be mentioned is the exponential growth in bid documentation which has occurred in recent years. It is by no means uncommon for as many as twenty ancillary documents to be issued to bidders along with the actual contract document. This documentation can include various studies which have been carried out relating to the project, environmental impact assessments, collateral warranty forms and the like. In addition, technical specifications have become more and more long-winded, even on design and build contracts where the successful bidder carries out his own design against performance requirements. Tenderers are expected to read this extraneous documentation and allow for it when preparing their bid, resulting inevitably in large numbers of queries during the tender period, the responses to which are then incorporated in the contract adding a further layer of documentation.

It is no coincidence that proliferation of documentation has coincided with the disappearance of hard copies of contracts. It is so easy for an employer or his consultant to simply attach a related study or specialist assessment already in electronic form when the documents themselves are issued electronically, rather than taking the trouble to extract what is relevant for incorporation in the contract. As an extreme example of how things have changed, I have copies of documents for works in Peterhead Harbour which were issued to bidders in 1878. The works involved rock excavation, strengthening and extending of quay walls, jetty construction and a seawall. They comprised:

Specification—8 pages

General Conditions—5 pages

Bill of Quantities—1 page

Tender—1 page

Reading these simple documents, it seems the scope of works was considerable, perhaps with a value of £50 million at present day prices. I can say that, from a knowledge of Peterhead Harbour, the works which were carried out at that time endured very well for nearly 140 years until the advent of larger fishing vessels required further major expansion of the facilities. Quality certainly did not suffer from having a simple, short contract. In management of construction the most important factors are an appropriate design and clarity which are not served by thousands of pages of waffle.

The Future of Civil Engineering

Throughout the twentieth century the work of civil engineers underpinned the massive social developments which took place all over the world, despite interruption by two world wars, and infrastructure projects became progressively larger over the course of the century. Apart from Brooklyn Bridge and Forth Bridge with spans of around 500 metres, few bridges in 1900 had spans of more than 25 metres, yet by the year 2000, spans of over 1,500 metres were commonplace. By linking bridges and tunnels in major transport schemes, large rivers and sea channels which had been barriers between regions were crossed. During the 100 years of the twentieth century new ports were built, first in natural harbours, and then, as world trade increased and vessels became larger requiring deeper water at their berths, in less suitable locations with high wave exposure or on poor ground, all difficulties being overcome by engineering ingenuity. So long as continued expansion of human activities was considered desirable, civil engineers were to the fore facilitating the expansion.

But today, well into the twenty-first century, we stand at an uncertain juncture. The ICE journals I read are full of papers on sustainability, and carbon neutral construction of one form or another. Consultants and contractors publicise their credentials on how they achieve or aspire to achieve this goal and there is less emphasis on engineering excellence. Few of the papers on this subject can match the authoritative elegance of the papers on civil engineering projects I read when I was a graduate member over fifty years ago and which encouraged me to seek corporate membership. In keeping with the thrust of many of the current papers in the journals, the ICE has sought to widen its membership base, with a focus on environmental disciplines and has even opened its membership to environmental specialists who are not engineers in the accepted sense of the word. If the aim is to ensure environmental issues receive proper recognition, which of course, they must, tinkering with membership requirements is unnecessary. A good engineer is used to managing projects which involve disciplines other than civil engineering, such as architecture, electrical installation and mechanical plant and equipment. One of the strengths of a civil engineer is, or should be, the ability to view projects holistically and work "from the general to the particular", as was often said, which is an attribute not always apparent in non-engineers and indeed often lacking in engineers of other disciplines. Most civil engineers I knew had a fundamental respect for the environment and would ensure projects caused as little lasting damage to it as possible. With this background we should be able to design and manage projects sustainably and recognise where specialist environmental help is needed. Putting environmentalists in positions of management, however, lets the tail wag the dog and may result in contrived project requirements often without any tangible environmental benefit as I saw on some aspects of the IWMF.

But in any case, this is just window-dressing and will have little effect on the main problem of CO_2 emissions. Most people now realise that much more fundamental action is needed if truly damaging changes to the planet's environment are to be avoided. Political will is lacking to reduce emissions in countries such as China and India, but with increasing prevalence of severe weather events and accelerating sea level rise, far-reaching steps to reduce emissions will eventually be agreed worldwide. These will almost certainly result in fewer large projects which are the mainstay of civil engineering with more emphasis on re-engineering of existing facilities to comply with strict emissions control. There may be less need for civil engineers overall, and instead a requirement for a wider and more far-seeing role for the civil engineers that remain. It is to be hoped the ICE recognises the importance of meeting this challenge in such a way that preserves the basic principles of civil engineering.

My Career in Context

I was always conscious I had shortcomings as an engineer although I carefully hid these from my peers throughout my career. Apart from a limited ability in mathematics which first became apparent in my university days, my main failing was carelessness, particularly in arithmetical work and a desire to get any task finished as quickly as possible to wrap up the job and move on to the next. I was also guilty of pretending to know more than I did, as most engineers do in my experience, not only of geotechnical engineering, some of the principles of which I only got to grips with quite late in my career, but even of structures, supposedly my specialty. To counter these failings, I learned to revisit calculations as a matter of course to pick up errors which were often present and to carry out repeated "sanity checks" on design values I had obtained.

On the credit side, a useful attribute I think I had was a flair for conceptual design in civil and structural engineering, and like a good rugby fly-half or midfielder in football, an ability to see several moves ahead and visualise how a design concept would develop. This may have compensated for my failings, especially when I took on more of a managerial role as I could get others to tackle the boring detailed calculations. And as with many other engineers, having gravitated into management, I found a lot of satisfaction in directing design teams on civil engineering projects to achieve a good outcome, not only for the client who had commissioned the project, but also for the contractor on the works. The key to this in my view is to make a design as "buildable" as possible, something which can only come from experience of designs which have not gone so well.

Finally, what of the projects I was responsible for which I have described in this memoir. The tender by the CNEPG/Paul Y joint venture for the IWMF on Shek Kwu Chau based on Mannings' design was not successful despite there being only two bidders left at the end, ours and a bid from a consortium led by Keppel of Singapore. The lack of nous exhibited by CNEPG in their technical submission which I struggled to edit may have been too off-putting for the client to countenance and it may be also that their bid price for the incinerator was too high. Nevertheless, if anyone is contemplating locating a public utility of some sort on an artificial island on soft ground in an inhospitable natural environment, Mannings and I have a ready-made design in a drawer just waiting for them.

Elsewhere in Hong Kong, a place where the pace of change is fast, the Trunk Road or Tolo Highway as it is now called still runs on its reclaimed embankment, though some of the bridges I designed have been demolished for construction of new interchanges. The trident blocks of Tsui Ping South estate on our site formation at Kai Liu are still used for public housing. In Bangkok, several other bridges have been constructed over the Chao Phraya River, but Dr Homberg's Rama IX Bridge with my approach viaducts leading to it is the most elegant of them all. Further down the river the dockyard is still an important base for the Thai Navy. The Kinabatangan Bridge is the only crossing of that large river in eastern Sabah and has provided a vital link for the people of the south In Lahad Datu and Tawau for over 30 years.

In Scotland, Carradale Breakwater, Trondra and Burra Bridges and Jedfoot Bridge are still in service, very much as they were when constructed in the 1960s and Ballachulish Bridge still frames the Narrows. Sullom Voe with its reconstructed dolphins continues to operate as the UK's largest oil export terminal and the Smith Quay at Peterhead helps PPA to go from strength to strength commercially, contributing substantially to the local economy of north-east Scotland. In London, construction of a set of reverse head mitre gates at Tilbury Docks as part of an upgrade of the entrance lock has rendered our floodgate redundant, but the gate protected the docks for forty years and was considered such good value that we were given a commission to design a similar type of floodgate at the entrance to Royal Docks further up the Thames.

Most of the works I have described and many others I was responsible for which do not feature in this memoir, in India, Malaysia, Indonesia and West Africa, are still in existence functioning satisfactorily, I hope, and have played their part in the economic development of countries and regions where I worked. If these places gained something from me, I gained in return from the experiences I had in them giving me a perspective of the world sometimes lacking in those who have not worked abroad. For me, the picture of the civil engineer my father painted in my mind when I was still at school came alive under the skies of Hong Kong and SE Asia, in the forests of Borneo and the bush of Northern Nigeria. I can take some satisfaction therefore that I tried to live up to the perception of civil engineering that he had from his own travels which led him to recommend it to me as a career.

Photographs

1. Collapsed box girder at Milford Haven
2. Ballachulish Bridge
3. Tilbury Docks Floodgate
4. Northern part of the Trunk Road in 1985
5. Treatment Works Bridge
6. Bridges at Island House Interchange
7. Kinabatangan Bridge
8. The summit of Mount Kinabalu
9. Thai Navy Dockyard at Pom Prachul
10. Rama IX Bridge, Bangkok
11. Rama IX Bridge Bangkok-side Approach Viaduct
12. Qandil Bridge
13. Peterhead Harbour
14. Smith Quay schematic structure drawing
15. Newly erected box girder for Smith Quay.
16. Precast concrete beams for Smith Quay deck
17. Sullom Voe oil jetties
18. Shear-leg barge at Sullom Voe carrying new dolphin skirt
19. Fitting of rubber cone fender system
20. Newly vulcanised rubber cone in factory
21. The IWMF

1. Collapsed Box Girder at Milford Haven

The collapse occurred in June 1972 while the box girder superstructure was being erected. The bridge had been designed by Freeman Fox & Partners who were supervising construction. At the time of the collapse the box girder was cantilevering over one of the piers by 70m and a deck segment weighing 150 ton was being moved along it ready to be bolted into position at the end of the cantilever. Forces on the box were probably in excess of anything which would be experienced in service. The cause of the failure was buckling of the diaphragm within the box over the pier. It was inadequately stiffened for the high reaction forces at the pier during erection. Four workers were killed in the collapse and five were injured. The incident had far-reaching consequences for bridge design in the UK and led to the introduction of the Interim Design and Workmanship Rules for steel girder bridges (the IDWR) and ultimately to the drafting of the comprehensive British bridge design code BS 5400.

2. Ballachulish Bridge

The bridge is viewed from the north side of the tidal narrows at Ballachulish looking west. As can be seen the truss chords are steel boxes and the web members (the verticals and diagonals) are open section steel plate girders. The bridge is continuous over the piers where the depth between top and bottom chords is increased to cater for higher global bending effects in the trusses at the supports. All joints between chords and web members are made with high strength friction grip bolts. The bottom chords sit on 3000 ton "pot" bearings on the pier tops which act like hinges. Pot bearings are also provided at the base of the piers and the bridge is anchored at the south abutment where the short approach span is tied down to a rock outcrop. After nearly 50 years' service the bridge is still in good condition.

3. Tilbury Docks Floodgate

The Western Entrance Lock at Tilbury Docks with the floodgate in its parked position on the east side of the lock entrance. The lead-in jetty at the entrance is in the foreground.

Commissioning of the floodgate. The frame has been launched across the entrance and the gate is being lowered into the lock for the first time.

4. Northern Part of the Trunk Road In 1985

This oblique aerial view is from the Tai Po end of the Trunk Road looking south-east shortly after opening of the road. The wooded area in the left foreground is at Island House, the residence of the secretary for the New Territories, at the time David Akers-Jones. The finger pier in the middle distance was constructed for the police station at Tai Po Kau which can be seen between the south-bound carriageway of the road and the shoreline. Ma On Shan is in the distance on the far side of Tolo Harbour.

5. Treatment Works Bridge

In this photograph looking north up the Trunk Road, construction of Treatment Works Bridge is almost finished with only surfacing of the south ramp to be completed. The prestressed concrete bridge crosses six lanes of the Trunk Road carriageways in a single span of over 45 metres. The anchored wall retaining the KCR embankment can be seen alongside the north-bound carriageway with a train rounding the curve close to the bridge ramp. At the time the photograph was taken landscaping and planting had yet to be commenced.

6. Bridges at Island House Interchange

Island House Interchange connects the Trunk Road to Tai Po at the northern end of the route where it swings west of the town towards Fanling. The photograph was taken on the day of the opening of the Trunk Road in 1985 before traffic was allowed on the road. The two bridges in the photograph are on curved alignments and were not suited to straight precast concrete post-tensioned beams which were used for most of the other bridges. The nearer bridge is a four-span structure with a flat slab reinforced concrete deck on circular piers, while second bridge is a two-span prestressed concrete box girder with a single pier in the central reserve.

7. Kinabatangan Bridge

Kinabatangan Bridge in 2018 over thirty years after it was constructed showing the main span of 144m across the river. The bridge is like Ballachulish Bridge in concept, with the bridge deck supported on sliding bearings on the cross beams between the trusses. However, in Kinabatangan Bridge, the trusses are in a classic Warren girder form with no vertical members and with a constant depth between the top and bottom chords throughout.

8. The Summit of Mount Kinabalu

The photograph was taken about one hour after sunrise as we started our descent from the summit. Clouds from the overnight thunderstorms have rolled back as the sun rose over the Sulu Sea to the east with the western side of Low's Peak still in deep shadow.

9. Thai Navy Dockyard at Pom Prachul

The two dry docks are to the right of the photograph, accessed off the semi-tidal basin. Both dry docks are in use with the gates closed and vessels in the docks undergoing maintenance. One of the frigates purchased from Britain can be seen berthed on the riverfront jetty.

10. Rama IX Bridge, Bangkok

The Rama IX Bridge in the final stages of construction in 1988, seen from the Thonburi side of the Chao Phraya River. Part of the Approach Viaduct on the Bangkok side can also be seen. With its main span over the river of 450m the Rama IX Bridge was the largest cable-stayed bridge in the world at that time. Now there are many cable-stayed bridges with spans over 1,000m, especially in China. The bridge was designed by Dr Helmut Homberg, for the consortium led by PFP. The elegant simplicity of its structure speaks for itself and shows that when a structure is well designed for its function, in this case to carry a six-lane highway over a river, it has no need for unnecessary embellishments or contrived structural form.

11. Rama IX Bridge Bangkok-Side Approach Viaduct

The northbound and southbound carriageways on the Rama IX Bridge approach viaducts are on separate structures with a gap of 1.0m between them. There were four separate bridge decks to construct therefore, two on either side of the river. The contractor provided two erection gantries, one for the Bangkok side and one for the Thonburi side. The first deck structure on the Bangkok side is shown on the left in a partially constructed condition with the erection gantry advancing up the approach to the main bridge. The three launching noses of the gantry can be seen reaching out towards the next pier support.

The photograph on the right shows the gantry with the leading ends of its box girders supported on the inclined "Pisa" tower devised by the contractor Maeda. Support on the "Pisa" tower 9m out from the main bridge pier (the Junction Pier) enabled the launching nose trusses to be dismantled in segments as the gantry approached the steel box girder of the main bridge. The gantry itself was then lowered to the ground using special support frames on the bridge deck and reused for the second bridge deck on the Bangkok side.

12. Qandil Bridge

The newly constructed Bailey bridge seen from the south bank of the river Zab just before I carried out my inspection. Contrary to the American air photo, there is no distortion of the structure which is true to line and level. The main span is a triple-double bridge and the side span beyond it a double-single. On the far bank of the river the downstream wingwall of the abutment had failed under earth pressure and the backfill had then washed out. Using additional bridging units, the resourceful Kurdistan contractor erected a side span over the gap which had been created, turning the north abutment into a free-standing pier. A queue of relief trucks can be seen extending back for some distance on the north bank. When I completed my inspection and had two heavily loaded vehicles traverse the bridge in a simple load test, the contractor and I waved the waiting trucks through to cross one at a time and the aid convoy was able to deliver much needed food and fuel to the south and east of Iraqi Kurdistan.

13. Peterhead Harbour

This oblique aerial photograph is taken looking due west into Peterhead Harbour before the Smith Quay was constructed off the so-called Smith Embankment in the area indicated on the photograph. It will be seen that this area got some wave protection from the main outer breakwaters and the Albert Quay Breakwater in the harbour itself. However, it still had considerable exposure to waves passing through the harbour entrance in storms from a south-easterly direction which often occur in winter. The existing inner breakwater therefore had to be extended to provide adequate protection if a new quay was to be constructed in that location.

14. Smith Quay Schematic Structure Drawing

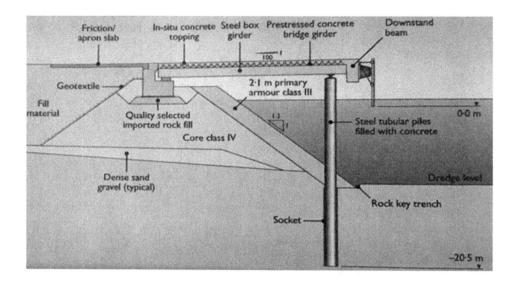

This schematic cross section through the quay shows the deck with its box girders, prestressed concrete bridge beams and in situ concrete topping making up a composite structure for live loading. Also shown is the revetment under the quay and the line of support piles at the toe of the revetment slope. The clear span of the deck over the revetment from the bank seat abutment to the line of pile supports is 24m. The deep beam at the front of the quay deck is formed by a steel box filled with concrete which is designed for direct impact on the beam by small fishing vessels when drawing their nets up on to the deck. Larger pelagic and general cargo vessels berth against rubber cone fenders which are generally located 10m apart in line with the steel box girders.

15. Newly Erected Box Girder for Smith Quay

In this photograph, the first box girder has just been erected. It is spanning 24m from the bank seat abutment to the rocker bearing on the pile top. Shear connector hoops can be seen on the top of the girder. Special hoops were used as the horizontal shear forces between the steel girders and the concrete deck are too high for standard stud connectors. As soon as possible after erecting this girder the adjacent girder was placed on its supports and the two girders were connected at the front ends by a segment of a steel down stand beam. Once this was done the pair of girders were completely stable and could be left while erection proceeded with the next pair.

16. Precast Concrete Beams for Smith Quay Deck

After erecting the steel box girders on the pile top bearings, the precast prestressed concrete bridge beams were placed between the top flanges of the girders. Once these bridge beams were in position a stable working surface was created and fixing of deck reinforcement could commence. Holes in the precast beam webs through which steel reinforcement was threaded can be seen. With the modular nature of the deck system construction was rapid and 120m of deck together with a berthing dolphin and connecting catwalk were completed within three months of erecting the first box girder.

17. Sullom Voe Oil Jetties

This photograph shows the oil terminal jetties at Sullom Voe shortly after the opening of the terminal by the Queen in 1980. The view looks north up the main approach to the terminal for oil tankers. The jetties are numbered 1 to 4 from south to north. The tankers in the photograph are lying against the berthing dolphins of Jetties 1, 2 and 3 with the mooring dolphins visible behind them. Work on replacing fender skirts and fenders described in the memoir was carried out on the berthing dolphins of Jetties 2 and 3, which are the centre two jetties in the photograph.

18. Shear-Leg Barge at Sullom Voe Carrying New Dolphin Skirt

The Norwegian shear-leg barge *Eide Lift* was hired by the Sullom Voe maintenance contractor to install the new dolphin skirts and cone fenders at Jetties 2 and 3. Although the barge owner in Norway quoted a fixed price for the North Sea crossing from Stavanger to Shetland, there was a charge on site in the oil terminal of £8,000 a day. Installation of the skirts and new fenders had to be carefully planned therefore to keep time on site of the barge as short as possible to minimise the overall cost of the installation. In planning the operations advantage also had to be taken of fine weather windows as there was a wind speed limit of 15knots for lifting operations.

19. Fitting of Rubber Cone Fender System

A new cone fender and its low-friction front facing panel being lifted into position at one of the berthing dolphins of Jetty 2. The new steel skirt has already been fitted on the front of the dolphin and its large prestressed fixing bolts can be seen near the top of the skirt. Once the cone fender has been bolted on to the skirt, the restraining chains (seen hanging loose in the photograph) will be connected to the front facing panel and tightened to hold it in position.

20. Newly Vulcanised Rubber Cone in Factory

Rubber cone fenders for the replacement dolphin skirts at Sullom Voe were manufactured in a factory in Qingdao in China. A total of 8 cones for the dolphins of Jetties 2 and 3 were manufactured in the factory where the workmanship was of the highest order. The photograph shows one of the cones which has just been removed from its vulcanising mould. This is a 2.0m cone which was the largest size available in 2015 when the cones were being made. The author spent three days in the factory witnessing loading tests on the four cones for the dolphins in a 5,000t press and viewing the various other quality control procedures which we had specified.

21. The IWMF

In this artist's impression, the IWMF is seen from the south-east, on the artificial island with the thickly wooded island of Shek Kwu Chau on the right. The incinerator with its 200m high flues sits in the centre of the 14 hectare-reclamation, surrounded by ancillary buildings. The 500m length of the seawall along the south of the artificial island protects the facility from waves which can reach 8m significant height in severe typhoons. To the west, the seawall extends into a main breakwater, also of 500m length, in the curved alignment stipulated in the specification for the works. Augmented by a secondary, detached breakwater, this provides protection for a harbour and unloading quay for the vessels transporting solid waste to the incinerator from refuse transfer stations located throughout Hong Kong.

CPSIA information can be obtained
at www.ICGtesting.com
Printed in the USA
BVHW052307290123
657301BV00003B/158